Thermal Performance of Membrane Distillation

Thermal Performance of Membrane Distillation

Editor

Alessandra Criscuoli

MDPI • Basel • Beijing • Wuhan • Barcelona • Belgrade • Manchester • Tokyo • Cluj • Tianjin

Editor
Alessandra Criscuoli
Istituto per la Tecnologia delle
Membrane (CNR-ITM)
Consiglio Nazionale delle
Ricerche
Rende (CS)
Italy

Editorial Office
MDPI
St. Alban-Anlage 66
4052 Basel, Switzerland

This is a reprint of articles from the Special Issue published online in the open access journal *Energies* (ISSN 1996-1073) (available at: https://www.mdpi.com/journal/energies/special_issues/Thermal_Performance_of_Membrane_Distillation).

For citation purposes, cite each article independently as indicated on the article page online and as indicated below:

LastName, A.A.; LastName, B.B.; LastName, C.C. Article Title. *Journal Name* **Year**, *Volume Number*, Page Range.

ISBN 978-3-0365-7768-5 (Hbk)
ISBN 978-3-0365-7769-2 (PDF)

Contents

About the Editor . . . . **vii**

Preface to "Thermal Performance of Membrane Distillation" . . . . **ix**

Peter M. Hylle, Jeppe T. Falden, Jeppe L. Rauff, Philip Rasmussen, Mads Moltzen-Juul, Maja L. Trudslev, et al.
Determination of Heat and Mass Transport Correlations for Hollow Membrane Distillation Modules
Reprinted from: *Energies* **2023**, *16*, 3447, doi:10.3390/en16083447 . . . . **1**

Alessandra Criscuoli and Maria Concetta Carnevale
Localized Heating to Improve the Thermal Efficiency of Membrane Distillation Systems
Reprinted from: *Energies* **2022**, *15*, 5990, doi:10.3390/en15165990 . . . . **17**

Marcello Pagliero, Marina Alloisio, Camilla Costa, Raffaella Firpo, Ermias Ararsa Mideksa and Antonio Comite
Carbon Black/Polyvinylidene Fluoride Nanocomposite Membranes for Direct Solar Distillation
Reprinted from: *Energies* **2022**, *15*, 740, doi:10.3390/en15030740 . . . . **35**

Marek Gryta
The Application of Open Capillary Modules for Sweeping Gas Membrane Distillation
Reprinted from: *Energies* **2022**, *15*, 1454, doi:10.3390/en15041454 . . . . **49**

Ravi Koirala, Xing Zhang, Eliza Rupakheti, Kiao Inthavong, Abhijit Date
Performance Study of Eductor with Finite Secondary Source for Membrane Distillation
Reprinted from: *Energies* **2022**, *15*, 8620, doi:10.3390/en15228620 . . . . **67**

Hosam Faqeha, Mohammed Bawahab, Quoc Linh Ve, Oranit Traisak, Ravi Koirala, Aliakbar Akbarzadeh and Abhijit Date
Solar Energy Driven Membrane Desalination: Experimental Heat Transfer Analysis
Reprinted from: *Energies* **2022**, *15*, 8051, doi:10.3390/en15218051 . . . . **87**

Alessandra Criscuoli
Thermal Performance of Integrated Direct Contact and Vacuum Membrane Distillation Units
Reprinted from: *Energies* **2021**, *14*, 7405, doi:10.3390/en14217405 . . . . **105**

Noah Yakah, Imtisal-e- Noor, Andrew Martin, Anthony Simons and Mahrokh Samavati
Wet Flue Gas Desulphurization (FGD) Wastewater Treatment Using Membrane Distillation
Reprinted from: *Energies* **2022**, *15*, 9439, doi:10.3390/en15249439 . . . . **117**

Mar Garcia Alvarez, Vida Sang Sefidi, Marine Beguin, Alexandre Collet, Raul Bahamonde Soria and Patricia Luis
Osmotic Membrane Distillation Crystallization of NaHCO$_3$
Reprinted from: *Energies* **2022**, *15*, 2682, doi:10.3390/en15072682 . . . . **129**

About the Editor

Alessandra Criscuoli

Alessandra Criscuoli (PhD) is a First Researcher at the Institute on Membrane Technology —National Research Council of Italy (CNR-ITM), Rende (CS) Italy. Expert in membrane operations, also as integrated systems, especially in membrane contactors applied to water and wastewater treatments and to water desalination. Participant, also as scientific responsible, in several national and international scientific projects on membrane development and applications. Co-author of one book on membrane contactors and of 120 contributions published in international peer-reviewed journals and as book chapters. Author of one comic book for children on membrane operations for water treatment. Co-editor of four books on membrane operations and water treatment and of journal special issues on membrane technology.

Preface to "Thermal Performance of Membrane Distillation"

Membrane distillation (MD) is a thermally driven membrane operation able to theoretically reject 100% of all non-volatiles contained in aqueous streams. It is based on the evaporation of the feed to be treated at the feed–membrane interface, the migration of the vapor/volatiles through the micropores, and the condensation and recovery of the permeated species at the distillate side. Membranes used are hydrophobic and microporous. The driving force of the process is the difference of partial pressure created across the membrane, and the temperature at the feed–membrane interface has been shown to have the greatest impact on the transmembrane flux. However, the temperature at the feed–membrane interface is usually lower than the feed bulk temperature because of temperature polarization phenomena, with a consequent decrease in the process efficiency. In addition, during MD, the feed is cooled inside the module, not only due to the evaporation but also due to the heat lost by conduction through the membrane matrix and the heat lost towards the environment. Therefore, the effective temperature for the evaporation is further reduced. This Special Issue focuses on the research efforts made to improve the thermal performance of MD, including the development of new module designs and heat recovery systems, the preparation of new types of membranes, the use of renewable energies and the integration with other membrane units. For instance, the analysis of heat and mass transport correlations, the development of membranes for localized heating, the design of new modules and condensation devices, the use of solar energy, the integration of different MD configurations and the thermal efficiency of MD in specific applications, are presented and discussed.

My most sincere thanks to all the Authors who contributed to the success of this Special Issue.

Alessandra Criscuoli
Editor

Article

Determination of Heat and Mass Transport Correlations for Hollow Membrane Distillation Modules

Peter M. Hylle, Jeppe T. Falden, Jeppe L. Rauff, Philip Rasmussen, Mads Moltzen-Juul, Maja L. Trudslev, Cejna Anna Quist-Jensen and Aamer Ali *

Department of Chemistry and Bioscience, Aalborg University, Fredrik Bajers Vej 7H, 9220 Aalborg, Denmark
* Correspondence: aa@bio.aau.dk

Abstract: Development and optimization of the membrane distillation (MD) process are strongly associated with better understanding of heat and mass transport across the membrane. The current state-of-the-art on heat and mass transport in MD greatly relies upon the use of various empirical correlations for the Nusselt number (Nu), tortuosity factor (τ), and thermal conductivity (κ_m) of the membrane. However, the current literature lacks investigations about finding the most representative combination of these three parameters for modeling transport phenomena in MD. In this study, we investigated 189 combinations of Nu, κ_m, and τ to assess their capability to predict the experimental flux and outlet temperatures of feed and permeate streams for hollow fiber MD modules. It was concluded that 31 out of 189 tested combinations could predict the experimental flux with reasonable accuracy ($R^2 > 0.95$). Most of the combinations capable of predicting the flux reasonably well could predict the feed outlet temperature well; however, the capability of the tested combinations to predict the permeate outlet temperatures was poor, and only 13 combinations reasonably predicted the experimental temperature. As a generally observed tendency, it was noted that in the best-performing models, most of the correlations used for the determination of κ_m were parallel models. The study also identified the best-performing combinations to simultaneously predict flux, feed, and permeate outlet temperatures. Thus, it was noted that the best model to simultaneously predict flux, feed, and permeate outlet temperatures consisted of the following correlations for τ, Nu, and κ_m: $= \frac{\varepsilon}{1-(1-\varepsilon)^{1/3}}$, $Nu = 0.13(Re)^{0.64}(Pr)^{0.38}$, $\kappa_m = (1-\varepsilon)\kappa_{pol} + \varepsilon\kappa_{air}$ where ε, Re, Pr, κ_{pol}, and κ_{air} represent membrane porosity, Reynolds number, Prandtl number, thermal conductivities of polymer and air, respectively.

Keywords: membrane distillation; modeling; Nusselt number; thermal conductivity; tortuosity factor

Citation: Hylle, P.M.; Falden, J.T.; Rauff, J.L.; Rasmussen, P.; Moltzen-Juul, M.; Trudslev, M.L.; Quist-Jensen, C.A.; Ali, A. Determination of Heat and Mass Transport Correlations for Hollow Membrane Distillation Modules. *Energies* **2023**, *16*, 3447. https://doi.org/10.3390/en16083447

Academic Editor: Chi-Ming Lai

Received: 7 March 2023
Revised: 28 March 2023
Accepted: 30 March 2023
Published: 14 April 2023

1. Introduction

Membrane distillation (MD) is a thermally driven process where the driving force is a vapor pressure difference created by a temperature difference across a porous hydrophobic membrane. MD can use low-grade heat from different sources, such as the sun, geothermal wells, and industrial processes, to produce ultra-fresh water [1–3]. MD is also an interesting candidate to achieve zero liquid discharge and crystallization from different solutions due to its ability to treat highly concentrated solutions, such as brine from desalination facilities [4–7]. The use of MD for simultaneous recovery of freshwater and minerals from different sources of impaired water makes it relevant to achieving sustainability and a circular economy [6,8].

MD can be operated in several configurations, including air gap, vacuum, sweep gas, and direct contact. Direct contact membrane distillation (DCMD) is the simplest configuration of the process in terms of the equipment and modules involved [9–11]. In DCMD, the membrane is in direct contact with the feed solution on one side and with the permeate on the other side. The driving force (i.e., vapor pressure difference) is induced by keeping the feed solution at a higher temperature than the permeate stream, creating a positive heat transfer through the membrane. Water and volatile compounds from the liquid feed evaporate, travel through the membrane pores, and are condensed at the membrane surface on the permeate side.

In DCMD, heat is transferred from the feed side to the permeate side due to the transport of water vapor through the membrane pores and conduction through the membrane [12,13]. As a result, the temperature at the membrane surface differs from its value in the bulk of the solution. The mass (vapor) flux across the membrane is directly linked with the difference in vapor pressures at the membrane surface on the feed and permeate sides, where the mass transfer coefficient of the membrane appears as a constant. The temperature difference between the bulk solution and membrane surface is known as temperature polarization, which decreases the effective driving force across the membrane and results in a reduction in transmembrane vapor flux [13]. The mass transfer coefficient of the membrane is a function of membrane properties including pore size, overall porosity, thickness, and the τ which, depending upon the membrane pore size and mean free path of the water vapor, can be calculated according to different models [3]. Determination of the vapor pressure at the surface requires knowledge of the membrane surface temperatures, which are linked with the bulk temperatures through heat transfer coefficients [14]. The thermal conductivity of the membrane (κ_m) affects the heat conducted across the membrane and therefore directly influences the total heat transport across the membrane and hence the temperature at the membrane surface [15]. Thus, the determination of flux is associated with the calculation of surface temperatures and the mass transfer coefficient of the membrane.

Understanding heat and mass transport in MD is important to design, improve, and optimize the process and module design [3,16–18]. Numerous semi-empirical correlations have been proposed to calculate the Nusselt number (Nu) for heat transfer coefficient in MD channels (see Section 2.3) [13,19]. The ultimate selection of the correlation for the Nu for a given fluid is a function of the applied hydrodynamics and the system configuration (e.g., flat sheet, hollow fiber). Likewise, Nu, several correlations have been proposed to describe the κ_m, including the parallel resistance model, the series resistance model, and the Maxwell I model, as described in Section 2.3. In all these correlations, effective membrane thermal conductivity models account for the membrane porosity, the thermal conductivity of stagnant air within the pore, and the thermal conductivity of the membrane material. High κ_m decreases the temperature gradient across the membrane, which results in lower vapor flux [20]. For τ, which is inversely linked with the vapor flux, three different approaches have been adopted [21,22]: (i) use it as an adjustable parameter in the model; (ii) use a constant value (usually between 1 and 2, but occasionally greater than 2) for the τ; and (iii) use theoretical approaches to link the membrane porosity with the τ.

Despite their fundamental importance, heat and mass transport in MD are poorly understood [13,23,24]. The current state-of-the-art modeling of MD approaches has two major limitations regarding the use of various correlations for the Nu, κ_m, and τ of the membrane. Firstly, they compare the validity of various correlations proposed for any of the three parameters (Nu, τ, and κ_m) for a fixed combination of the other two parameters. In other words, the state-of-the-art approaches do not test the validity of various combinations of correlations for the Nu, κ_m, and τ of the membrane. For instance, in some studies for the determination of a suitable correlation for the Nu, it was assumed that the κ_m could be represented by the parallel model [13,25]. Phattaranawik et al. considered the suitability of κ_m correlations in their DCMD model but neglected the correlations for τ [26]. Kim et al. studied the effect of using eleven different correlations for the τ on the flux and concluded that the use of an inappropriate correlation can incorporate a significant error in the predicted flux [27]. However, the study was carried out by assuming that heat transport within the membrane and in feed and permeate channels can be described by using a fixed combination of correlations for κ_m and Nu. This approach is clearly very specific to the membrane and operating setup applied in each of the studies, and its validity for a broad set of membrane and module characteristics cannot be guaranteed. The second important limitation of the current state-of-the-art is that the validation of different correlations for Nu,

κ_m, and τ has been tested by comparing the theoretical and experimental values of vapor flux only [11,15,27–29]. The potential of these models to predict the outlet temperatures of feed and permeate, which are crucial to calculating thermal and cooling energy demand, respectively, is broadly neglected in the current literature.

The overall objective of the current study is to analyze the capability of various combinations of state-of-the-art correlations for Nu, κ_m, and τ to predict the experimental flux and outlet temperatures for hollow fiber membrane modules. The ultimate objective is to find the best-suited combination of Nu, κ_m, and τ to predict the experimental data (flux and outlet temperatures).

2. Materials and Methods

2.1. MD Test

Experimental analysis of DCMD has been conducted to validate the model predictions by using a polypropylene hollow fiber membrane from Membrana GmbH. The membrane has a porosity of 73%, a mean pore size of 0.2 μm, and a thickness of 450 μm. The membrane module, consisting of 19 hollow fibers with an effective length of 51 cm placed in a shell with a 2.1 cm internal diameter, was fabricated in the laboratory. The experiments were performed with pure water as feed and permeate circulating on the lumen and shell sides, respectively, of the hollow fiber membrane module operating in the countercurrent mode. The Reynolds numbers for the permeate and feed sides were in the ranges 130–300 and 200–1800, respectively. Inlet and outlet temperatures of the feed and permeate streams were measured by using thermocouples (TECPEL thermometers). The flux was measured by following the weight loss of the feed container as a function of experimental time. Circulation of hot and cold streams was achieved by using a peristaltic pump (Masterflex L/S). The temperatures of the feed and permeate streams were controlled by using heating (Grant) and cooling (Julabo 200F) systems, respectively. Detailed experimental design parameters can be found in Table 1, whereas a schematic illustration of the experimental setup can be found elsewhere [30–32]. The experimentation was designed to comprehend the effect of different operating parameters, including feed inlet temperature and feed and permeate flow rates, on the process performance.

Table 1. Experimental temperatures and flow rates used in the study.

No.	Feed Inlet Temperature (°C)	Permeate Inlet Temperature (°C)	Feed Flow Rate (Lh^{-1})	Permeate Flow Rate (Lh^{-1})
1	35	12	99	29
2	39	13	99	29
3	44	14	99	29
4	48	15	99	29
5	49	18	99	29
6	49	22	99	29
7	57	17	99	29
8	65	16	99	29
9	48	14	68	29
10	48	14	43	29
11	47	14	29	29
12	49	14	99	23
13	48	15	99	32
14	48	15	99	50

2.2. Model Development

Mass flux in MD can be described mathematically as follows:

$$J = B\left(P_{fm} - P_{pm}\right) \tag{1}$$

where P_{fm} *and* P_{pm} are the vapor pressures at the membrane surface on the feed and permeate sides, respectively. B is a characteristic membrane parameter, and different models can be used to determine it [3]. The ultimate selection of the model depends on the mean free path of the water vapor and the nominal pore size of the membrane. For PP hollow fiber membranes, the mean free path of water molecules and the nominal pore size are in the same order of magnitude [20,33], and thus the combined Knudsen- and molecular diffusion model is used to determine B (see Equation (2)).

$$B = \left[\frac{3\tau\delta_m}{2\varepsilon r} \left(\frac{\pi R T_m}{8M} \right)^{\frac{1}{2}} + \frac{\tau\delta_m P_{air} R T_m}{\varepsilon P_{tot} D M} \right]^{-1} \tag{2}$$

P_{air} is the air pressure, and P_{tot} is the total pressure inside the membrane [3]. τ can be described by different correlations, as detailed in Section 2.3. D is the vapor diffusivity and is affected by both the temperature and P_{tot}. An empirical correlation between P_{tot} and D is proposed in a study by Yun et al. [4] (see Equation (3)) and is used in our study.

$$P_{tot} D = 1.19 \cdot 10^{-4} \cdot T^{1.75} \tag{3}$$

To estimate the mass flux over the membrane, the difference in vapor pressure at the two membrane surfaces must be known. The vapor pressure is described as a function of temperature in the Antoine equation (see Equation (4)) [5]. Therefore, temperature on the membrane surfaces in the feed (T_{fm}) and permeate (T_{pm}) is a prerequisite for the calculation of P_{fm} and P_{pm}.

$$P = \exp\left(A_a - \frac{B_a}{T + C_a} \right) \tag{4}$$

Since mass flux is driven by a temperature gradient between the feed and permeate streams, an accurate calculation of the heat transfer across the membrane must be applied. Heat transfer in MD can be divided into three steps [6]:

1. Heat transfer from the feed bulk to the membrane surface with the rate Q_f;
2. Heat transfer across the membrane with the rate Q_m;
3. Heat transfer from the boundary layer to the bulk solution on the permeate side is at a rate of Q_p.

Both the heat transfer in the permeate and feed solutions are convection processes and thus dependent upon a convective heat transfer coefficient ($h_{f/p}$) and the temperature polarization according to Equations (5) and (6) [34]:

$$Q_f = h_f\left(T_f - T_{fm} \right) \tag{5}$$

$$Q_p = h_p\left(T_{pm} - T_p \right) \tag{6}$$

The heat transfer coefficient in the feed and permeate solution can be estimated from the following correlation between Nu, the thermal conductivity of the solution $\kappa_{p/f}$, and the equivalent diameter D_e of the channel (see Equation (7)).

$$h_{p/f} = \frac{Nu_{p/f} \cdot \kappa_{p/f}}{D_e} \tag{7}$$

The equivalent diameter corresponds to the diameter of the fibers in the solution on the lumen side. For the solution on the shell side, Equation (8) should be used due to the triangular arrangement of the fibers [7]. Furthermore, multiple empirical correlations have been proposed to determine Nu. Thus, to ensure the most accurate prediction of the flux, different Nu correlations are investigated in this study (see Section 2.3) [14].

$$D_e = \frac{3.44 \cdot P_t^2 - \pi \cdot d_{out}^2}{\pi \cdot d_{out}} \tag{8}$$

d_{out} is the outer diameter of the fibers, and P_t is the average distance between the centers of the fibers [7].

The heat transfer rate across the membrane is affected by the κ_m and the mass transfer across the membrane (see Equation (9)) [35].

$$Q_m = \frac{\kappa_m}{\delta_m}\left(T_{fm} - T_{pm}\right) + J\Delta H_v = h_c\left(T_{fm} - T_{pm}\right) + h_v\left(T_{fm} - T_{pm}\right) \tag{9}$$

where κ_m can be calculated according to different correlations relating the thermal conductivity of the polymer used as membrane material, κ_{pol} and the thermal conductivity of the air, κ_{air}, present in the pores, as detailed in Section 2.3 [6]. δ_m is the thickness of the membrane, and ΔH_v is the latent heat of water vapor, which is dependent on the temperature at the membrane surfaces (see Equation (10))

$$\Delta H_v = 1.7535\left(\frac{T_{fm} + T_{pm}}{2} + 2024.3\right) \tag{10}$$

Under the steady state conditions:

$$Q_f = Q_m = Q_p \tag{11}$$

Due to the equality of the different heat transfer rates at steady state, the following correlations between T_{fm}, T_{pm}, T_f, and T_p can be derived [3].

$$T_{fm} = T_f - (T_f - T_p)\frac{\frac{1}{h_f}}{\frac{1}{h_v+h_c} + \frac{1}{h_p} + \frac{1}{h_f}} \tag{12}$$

$$T_{pm} = T_p + (T_f - T_p)\frac{\frac{1}{h_p}}{\frac{1}{h_v+h_c} + \frac{1}{h_p} + \frac{1}{h_f}} \tag{13}$$

To calculate the temperature profile along the module, the energy balance along the fibers must be applied. This is done by dividing the system into (L/n) elements, where L is the total length of the system and n is the total number of elements. The energy difference between the entrance and exit on the feed side for the i-th element is equal to the amount of energy transferred across the membrane due to conduction and convection (see Equations (14) and (15)), with a corresponding correlation also applicable for the permeate stream [3]. Thus, the feed and permeate temperatures for element $i + 1$ along the fiber for the countercurrent configuration can be calculated as follows:

$$T_{f|i+1} = \frac{\dot{m}_f C_p T_{f|i} - \left(\frac{\kappa_m}{\delta_m}\left(T_{fm} - T_{pm}\right)dA + J\Delta H_v dA\right)}{\dot{m}_f C_{p|i+1}} \tag{14}$$

$$T_{p|i+1} = \frac{\dot{m}_p C_p T_{p|i} - \left(\frac{\kappa_m}{\delta_m}\left(T_{fm} - T_{pm}\right)dA + J\Delta H_v dA\right) + Q_e}{\dot{m}_p C_{p|i+1}} \tag{15}$$

Calculating temperature profiles along the module length allows for the determination of mass transfer both parallel to the flow and across the membrane in discrete steps. This is simply calculated as the difference of mass flux along the module and across the membrane of the previous element:

$$\dot{m}_f|_{i+1} = \dot{m}_f|_i - J_i dA \tag{16}$$

The heat transfer to the environment (Q_e in Equation (18)) is firstly governed by the convective heat transfer from the bulk solution on the shell side to the inner surface of the module with a rate Q_{inner}, and secondly by the conductive heat transfer across the module with a rate Q_{module}. Lastly, heat is transferred from the outer surface of the module to the environment via both convection with the rate Q_{outer} and radiation with the rate Q_r. The formula for the different rates is presented in Equation (17) [8].

$$\begin{aligned}
Q_{inner} &= h_{inner} A_{inner} (T_{bulk} - T_{s,inner}) \\
Q_{module} &= \frac{k_{module}}{\delta_{tube}} A_m (T_{s,inner} - T_{s,outer}) \\
Q_{outer} &= h_{outer} A_{outer} (T_{s,outer} - T_{air}) \\
Q_r &= A_{outer} C\epsilon (T_{s,outer}^4 - T_{air}^4)
\end{aligned} \tag{17}$$

At steady state, the following equality for the heat loss can be applied.

$$Q_e = Q_{in} = Q_{module} = Q_{out} + Q_r \tag{18}$$

2.3. Variable Parameters

Three parameters used for the modeling computations are varied, i.e., τ, κ_m, and Nu. Nine correlations for the Nu, seven for τ, and three for the κ_m are tested. The specific details about the correlations used are presented in Table 2. All possible combinations of the considered correlations are tested, i.e., 189 combinations in total. Nu correlations are empirically determined and depend on various dimensionless parameters, module length, and equivalent diameter. The correlations presented in the table are valid in the laminar flow regime, i.e., $Re < 2100$. Considered correlations for κ_m are the parallel resistance model, the series resistance model, and the Maxwell I model [8].

Table 2. Correlations of τ, Nu, and κ_m used in the DCMD computations.

Tortuosity (τ)	
$\tau_1 = \frac{(2-\varepsilon)^2}{\varepsilon}$	[27]
$\tau_2 = \frac{\varepsilon}{1-(1-\varepsilon)^{1/3}}$	[28]
$\tau_3 = \frac{1}{\sqrt{\varepsilon}}$	[30]
$\tau_4 = \frac{1}{\varepsilon}$	[27]
$\tau_5 = \frac{(3-\varepsilon)}{2}$	[30]
$\tau_6 = \sqrt{1 - ln(\varepsilon/2)}$	[30]
$\tau_7 = \frac{\varepsilon}{1-(1-\varepsilon)^{2/3}}$	[30]
Nusselt number (Nu)	
$Nu_1 = 1.86\left(\frac{RePr}{L/D_e}\right)^{1/3}$	[31]
$Nu_2 = 4.36\left(\frac{0.036RePr(D_e/L)}{1 + 0.0011RePr(D_e/L)^{0.8}}\right)$	[31]
$Nu_3 = 0.13(Re)^{0.64}(Pr)^{0.37}$	[32]
$Nu_4 = 1.95\left(\frac{RePr}{L/D_e}\right)^{1/3}$	[33]
$Nu_5 = 0.097(Re)^{0.73}(Pr)^{0.13}$	[34]
$Nu_6 = 3.66 + \left(\frac{0.104RePr(D_e/L)}{1 + 0.106RePr(D_e/L)^{0.8}}\right)$	[33]
$Nu_7 = 1.62(RePrD_e/L)^{0.33}$	[35]
$Nu_8 = 4.36 + \frac{0.023Pe/(L/D_e)}{1 + 0.0012Pe/(L/D_e)}$	[36]
$Nu_9 = 4.364 + \frac{0.02633}{(L/(D_ePe)^{0.506})\exp(41L/(D_ePe))}\left(\frac{Pr}{Pr_w}\right)^k$ $k = 0.20$—feed, $k = 0.19$—permeate	[36]

Table 2. *Cont.*

Membrane thermal conductivity (κ_m)	
$\kappa_{m,1} = (1 - \varepsilon)\kappa_{pol} + \varepsilon\kappa_{air}$	[37]
$\kappa_{m,2} = \left(\frac{\varepsilon}{\kappa_{air}} + \frac{1-\varepsilon}{\kappa_{pol}} \right)^{-1}$	[37]
$\kappa_{m,3} = \kappa_{air}\left(\frac{1 + (1-\varepsilon)2\beta_{pol-air}}{1-(1-\varepsilon)2\beta_{pol-air}} \right) \beta_{pol-air} = (\kappa_{pol} - \kappa_{air})/(\kappa_{pol} + 2\kappa_{air})$	[37]

3. Results and Discussion

3.1. Experimental Mass Fluxes and Outlet Temperatures

Details of experimental input (inlet temperatures and flow rates) and output (outlet temperatures and fluxes) parameters are provided in Table 3. It is evident from the table that the flux increases with an increase in feed temperatures and flow rates. Experiments 1–8 indicate that flux increases exponentially from 1.05 to 6.07 kg/m^2.h by increasing the feed temperature from 35 to 65 °C at a constant feed and permeate flow rate. This agrees with the exponential dependence of flux on temperature in MD reported in the literature [3,36]. The corresponding feed outlet temperature also increases, from 34 to 60 °C. A similar trend is also observed for the permeate outlet temperature. Experiments 9–11 were aimed at investigating the effect of feed flow rate on transmembrane flux and outlet temperatures. It is evident from the corresponding data that the flux decreases with the feed flow rate. This is a direct consequence of increased temperature polarization and feed temperature drop along the membrane module, as evident from the corresponding T_{fout} data. The observed trend again corresponds to the observations reported in the literature [37]. Experiments 12–14 were carried out to explore the effect of permeate flow rates on the mass flux and temperature drops along the module. As evident from the corresponding flux data reported in Table 3, flux exhibits a very weak dependence upon the permeate flow rate. Both feed and permeate outlet temperatures drop slightly with an increase in permeate flow rate, which was expected due to improved heat transfer on the permeate side and a shorter residence time for the permeate stream inside the module.

Table 3. Experimentally measured inlet and outlet temperatures and trans-membrane flux at known feed and permeate temperatures and flow rates.

No.	$T_{f,in}$ [°C]	$T_{f,out}$ [°C]	$T_{p,in}$ [°C]	$T_{p,out}$ [°C]	Q_f [L/h]	Q_p [L/h]	J [kg m^{-2}h^{-1}]
1	35	34	12	18	99	29	1.05
2	39	38	13	20	99	29	1.35
3	44	42	14	21	99	29	1.84
4	48	46	15	24	99	29	2.35
5	49	47	18	27	99	29	2.27
6	49	47	22	29	99	29	2.38
7	57	54	17	29	99	29	3.76
8	65	60	16	32	99	29	6.07
9	48	42	14	23	68	29	2.13
10	48	42	14	23	43	29	2.15
11	47	38	14	22	29	29	1.84
12	49	47	14	25	99	23	2.45
13	48	47	15	23	99	32	2.43
14	48	46	15	22	99	50	2.48

3.2. Evaluation of the Model Predictions

All possible combinations of Nu, κ_m, and τ shown in Figure 1 have been evaluated to assess their ability to predict the experimental values of the average flux across the membrane and the outlet temperatures $T_{f,out}$ and $T_{p,out}$. The accuracy of the results is determined by the R^2-value of the fit to the function $y = x$ of the experimental (y) and model prediction (x) data. Detailed results are shown in Table A1 in the Appendix A, which indicates that the R^2 values for flux and outlet temperatures predicted by each of the combinations differ widely. Thus, R^2-values for J fall in the interval $-2.866 < R^2 < 0.989$, while R^2 values for $T_{f,out}$ and $T_{p,out}$ lie in the intervals $0.891 < R^2 < 0.977$ and $-0.529 < R^2 < 0.969$, respectively. The variation of R^2-values across all models is noteworthy, as it indicates that model predictions are significantly affected by the combination of the adjustable parameters used.

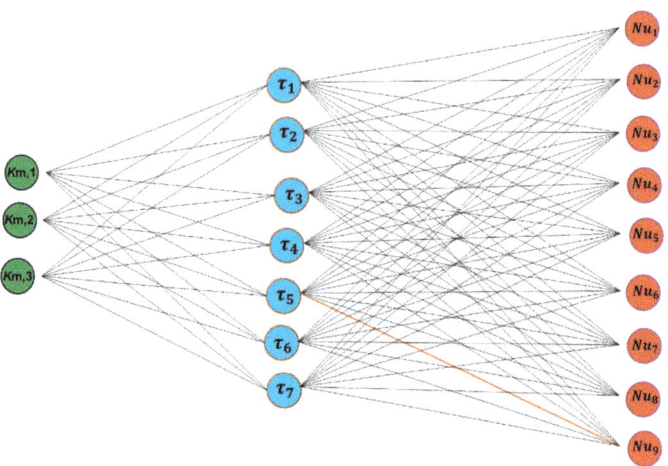

Figure 1. An illustration of different combinations of k_m, τ, and Nu tested in the study. The numbering for each correlation is according to Table 2.

A statistical analysis of the data has been provided in Figure 2. It is evident from Figure 2 that only 31 out of 189 combinations could predict the flux with accuracy exceeding $R^2 > 0.9$. This emphasizes the significance of using the appropriate combination of tortuosity, heat transfer, and κ_m correlations in DCMD modeling. Furthermore, the variation and the number of low R^2-values suggest that many combinations yield inaccurate predictions. Thus, evaluation of the prediction performance of a given model may have to account for the relative model performance rather than the absolute performance. The number of combinations that could predict $T_{f,out}$ with $R^2 > 0.9$ was the highest (180). On the other hand, only 13 combinations could predict $T_{p,out}$ with reasonable accuracy ($R^2 > 0.9$).

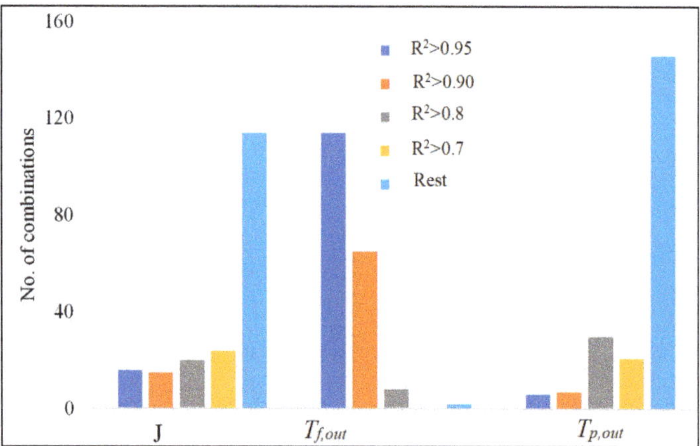

Figure 2. Statistical analysis of different combinations of Nu, τ, and κ_m used to predict the experimental parameters.

3.3. Selection of the Best Fitting Model

The selection of the best-fitting model in this study is based on the highest sum of the R^2-values for J, $T_{f,out}$, and $T_{p,out}$. The combinations of the three adjustable parameters τ, Nu, and κ_m with the best overall prediction performance based on the conditions set in this study are reported in Table 4 along with the corresponding R^2-values. The accuracy of the fit to flux, feed, and permeate outlet temperatures is represented with $R^2(J)$, $R^2(T_{f,out})$, and $R^2(T_{p,out})$, respectively. R^2_{tot} is the average of $R^2(J)$, $R^2(T_{f,out})$, and $R^2(T_{p,out})$.

Table 4. R^2-values and correlation numbers (numbers for τ, Nu, and κ_m correspond to Table 3) for the best-fitting model (based on fit for $T_{f,out}$, $T_{p,out}$, and J).

τ	Nu	κ_m	$R^2(J)$	$R^2(T_{f,out})$	$R^2(T_{p,out})$	R^2_{tot}
2	3	1	0.970	0.976	0.951	0.966
1	3	1	0.911	0.976	0.948	0.945
2	5	1	0.951	0.973	0.907	0.944
1	5	1	0.880	0.973	0.903	0.918
6	9	1	0.942	0.970	0.805	0.906

The model with the best predictability for all the parameters is the one that combines τ_2, Nu_3, and $\kappa_{m,1}$. The experimental values are plotted against the predicted values for this model in Figure 3 to confirm the model's performance. It is evident from the figure that the model predicts all three parameters well, but standard deviations vary between J, $T_{f,out}$, and $T_{p,out}$, where J-standard deviations are generally higher than those of $T_{p,out}$ and especially $T_{f,out}$.

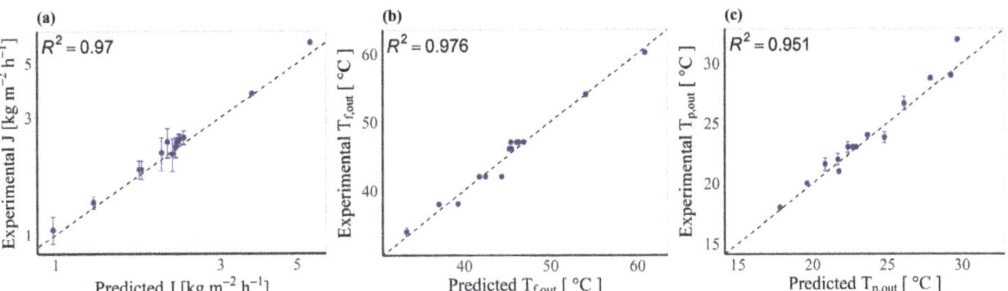

Figure 3. Model using τ_2, Nu_3, and $\kappa_{m,1}$. (**a**) Experimental versus predicted J (both axes are on a logarithmic scale of base 10); (**b**) experimental vs. predicted $T_{f,out}$; and (**c**) experimental vs. predicted $T_{p,out}$. Note that error bars in some cases are smaller than the points and, therefore, are not visible for some points.

One may note from Table A1 that the model with the best overall prediction ability does not have the best prediction ability when solely focusing on the flux. Therefore, it is necessary to investigate the best model when only taking flux prediction into account. Several combinations of τ, Nu, and κ_m can determine the flux accurately. The best five models for flux prediction are shown in Table 5 along with their corresponding R^2 values for their general predictions. It is evident from the table that even though a given model can predict the flux well, this does not necessarily entail that it is also able to predict the outlet temperatures well.

It is evident from Table 5 and Figure 4 that the best models to predict J can also predict feed outlet temperature reasonably well with R^2-values above 0.9; however, their capability to predict permeate outlet temperatures is poor, and the corresponding R^2-values fall in the range $-0.187 < R^2 < 0.758$. However, accurate prediction of outlet temperatures for permeate as well as the feed stream is important as these parameters play a significant role in the heating and cooling energy consumption of the DCMD process.

Table 5. R^2-values and correlation numbers (numbers for τ, Nu, and κ_m correspond to Table 3) for the best-fitting model for the flux.

τ	Nu	κ_m	$R^2(J)$	$R^2(T_{f,out})$	$R^2(T_{p,out})$	R^2_{tot}
2	3	3	0.989	0.965	0.758	0.904
1	3	2	0.986	0.916	−0.071	0.610
1	5	2	0.985	0.910	−0.187	0.569
2	5	3	0.984	0.959	0.657	0.866
2	5	2	0.978	0.911	−0.154	0.578

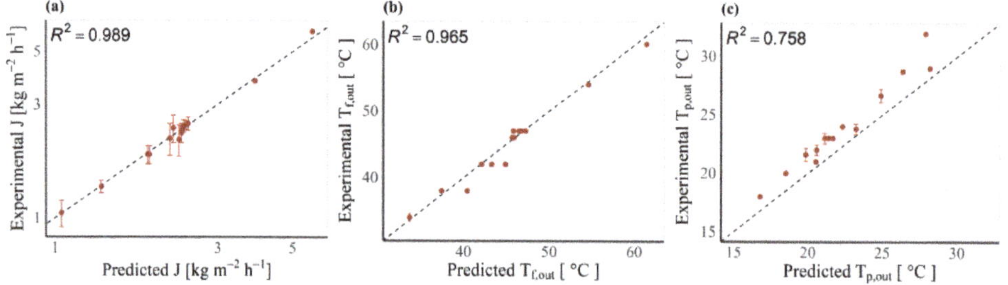

Figure 4. Model using τ_2, Nu_3, and $\kappa_{m,3}$. (**a**) Experimental versus predicted J (both axes are on a logarithmic scale of base 10); (**b**) experimental vs. predicted $T_{f,out}$; and (**c**) experimental vs. predicted $T_{p,out}$. Note that the error bars in some cases are smaller than the points.

4. Observed Tendencies

All different combinations of adjustable parameters and the corresponding R^2 values are shown in Table A1 in the Appendix A. The results have been reported in decreasing order with respect to R_{tot}^2. From Table A1, it is evident that R_{tot}^2 values vary significantly for different combinations, and even though it is not possible to clearly evaluate the prediction accuracy for every correlation, some tendencies can be observed.

It is noteworthy that among the best-performing models, the most commonly used correlation for the determination of κ_m is $\kappa_{m,1}$. This indicates that $\kappa_{m,1}$ can describe the heat transfer across the membrane more accurately than the remaining correlations. Contrarily, most of the models with poor overall prediction performance use $\kappa_{m,2}$ correlation. Models based on $\kappa_{m,2}$ especially show poor prediction of the outlet temperatures, suggesting that κ_{air} and κ_{pol} are not weighted appropriately in the series resistance model ($\kappa_{m,2}$) and that the focus of this correlation may be flux prediction.

A similar tendency is observed when examining tortuosity correlations, where τ_2, when paired with the appropriate correlations for Nu and κ_m, best predicts the experimental outputs. The poorest prediction of the correlations for τ follows the order τ_5, τ_3, and τ_7. Generally, a tendency is observed where a lower value of τ corresponds to a lower prediction ability, except for τ_1 and τ_2. This suggests that the optimal value of τ is around 2.0 for a membrane with high porosity, which is also reported in a study by Khayet et al. [3]. The value of τ has a direct impact on the permeability coefficient, B, which is used to determine the transmembrane flux. Higher values of τ entail lower B and thereby also lower predicted flux. This observation leads to the conclusion that most of the correlations used for τ tend to overestimate the flux.

When observing the Nu correlations, no clear tendency was observed. This is evident from the fact that Nu_3 and Nu_5 are present in both the models with the highest and lowest prediction accuracy. This might also suggest that the choice of the Nu correlation is of less significance than those of τ and κ_m. Furthermore, these observed tendencies for the correlations of τ, Nu, and κ_m emphasize the importance of making the correct choice of model for theoretical DCMD modeling.

5. Conclusions

In the present work, different combinations of state-of-the-art empirical correlations for the tortuosity factor, Nusselt number, and thermal conductivity of the membrane have been evaluated for their ability to predict outlet temperatures and flux in DCMD. In total, 189 combinations have been investigated, with varying results. Only 31 combinations were able to predict the experimental flux with reasonable accuracy ($R^2 > 0.95$) whereas 180 correlations could accurately predict the feed outlet temperature. The worst predictions were observed for the permeate outlet temperatures, where only 13 tested combinations could predict the experimental temperature with reasonable accuracy. Only five combinations could predict all three set parameters simultaneously (flux, feed, and permeate outlet temperatures) with reasonable accuracy. This highlights the importance of using an appropriate combination of the adjustable parameters when using a theoretical model to predict the performance of a DCMD setup. The model with the best ability to predict J, $T_{f,out}$, and $T_{p,out}$ consists of τ_2, Nu_3, and $\kappa_{m,1}$, where the numbering in subscript is according to Table 2. This model has an average $R^2 = 0.97$ for the fit of predicted values against experimental values.

Author Contributions: Conceptualization, A.A.; Methodology, P.M.H., J.T.F., J.L.R., P.R., M.M.-J., M.L.T. and A.A.; Software, P.M.H., J.T.F., J.L.R., P.R. and M.M.-J.; Validation, P.M.H., J.T.F., J.L.R., P.R., M.M.-J. and M.L.T.; Formal analysis, P.M.H., M.M.-J. and C.A.Q.-J.; Investigation, J.L.R. and P.R.; Writing—original draft, P.M.H., J.T.F., J.L.R., P.R. and M.M.-J.; Writing—review & editing, C.A.Q.-J. and A.A.; Supervision, C.A.Q.-J. and A.A.; Project administration, A.A.; Funding acquisition, C.A.Q.-J. All authors have read and agreed to the published version of the manuscript.

Funding: This research was funded by the Ministry of Foreign Affairs of Denmark (MFA) through the Danida Fellowship Centre (DFC) under Grant no. 20-M01AAU ("Membrane crystallization for water and mineral recovery").

Data Availability Statement: No data, in addition to what is reported in the manuscript, were created or analyzed in this study. Data sharing is not applicable to this article.

Acknowledgments: The authors are thankful to the Ministry of Foreign Affairs of Denmark for their financial support through the Danida Fellowship Centre for this work.

Conflicts of Interest: The authors declare no conflict of interest.

Abbreviations

Math symbols

τ	Tortuosity
κ	Thermal conductivity
κ_{pol}	Thermal conductivity of polymer
κ_{air}	Thermal conductivity of air
J	Mass flux
P	Vapour pressure
B	Membrane characteristic parameter
d	Nominal pore size
λ	Mean free path of water vapour molecules
K_n	Knudsen number
δ_m	Membrane thickness
ε	Membrane porosity
r	Pore radius
R	Gas constant
T	Temperature
D	Diffusivity of water vapour
M	Molecular weight
A_a, B_a, C_a	Antoine's equation coefficients
Q	Heat flux

h	Heat transfer coefficient
D_e	Equivalent diameter
d_{out}	Outer fiber diameter
P_t	Average distance between fibers
ΔH_v	Latent heat of water vapour
L	Module length
C_p	Isobaric heat capacity

Appendix A

Table A1. R^2 values for the correlation between experimental and theoretical data for J, $T_{f,out}$, $T_{f,out}$, and total average R^2. Model no. indicates the correlations used for the model in the following order: τ, Nu, and κ_m.

MODEL NO.	$R^2(J)$	$R^2(T_{f,out})$	$R^2(T_{p,out})$	R^2_{tot}
231	0.970	0.976	0.951	0.966
131	0.911	0.976	0.948	0.945
251	0.951	0.973	0.907	0.944
151	0.880	0.973	0.903	0.918
691	0.942	0.970	0.805	0.906
621	0.905	0.971	0.838	0.905
233	0.989	0.965	0.758	0.904
681	0.923	0.970	0.817	0.903
661	0.892	0.970	0.836	0.899
491	0.908	0.970	0.808	0.895
641	0.924	0.966	0.783	0.891
133	0.959	0.964	0.746	0.890
481	0.880	0.970	0.820	0.890
421	0.856	0.971	0.840	0.889
611	0.939	0.964	0.756	0.886
461	0.837	0.970	0.839	0.882
441	0.880	0.966	0.785	0.877
411	0.901	0.965	0.758	0.875
253	0.984	0.959	0.657	0.866
671	0.971	0.959	0.658	0.863
471	0.952	0.959	0.660	0.857
261	0.780	0.969	0.809	0.853
153	0.941	0.958	0.645	0.848
221	0.760	0.970	0.811	0.847
791	0.756	0.970	0.815	0.847
781	0.701	0.970	0.827	0.833
281	0.737	0.969	0.791	0.832
721	0.657	0.971	0.847	0.825
771	0.846	0.959	0.666	0.824
711	0.739	0.965	0.766	0.823
241	0.745	0.965	0.756	0.822
741	0.700	0.966	0.792	0.820
291	0.699	0.969	0.779	0.816
761	0.623	0.970	0.846	0.813
161	0.664	0.969	0.805	0.813
651	0.524	0.974	0.933	0.811
121	0.642	0.970	0.807	0.806
211	0.722	0.964	0.730	0.806
181	0.615	0.969	0.787	0.790
391	0.569	0.970	0.821	0.787
263	0.882	0.952	0.516	0.783
141	0.624	0.965	0.752	0.780
223	0.867	0.953	0.519	0.780
371	0.703	0.959	0.671	0.778
631	0.388	0.976	0.965	0.776

Table A1. *Cont.*

MODEL NO.	$R^2(J)$	$R^2(T_{f,out})$	$R^2(T_{p,out})$	R^2_{tot}
693	0.834	0.954	0.538	0.775
191	0.573	0.969	0.775	0.772
623	0.768	0.956	0.584	0.769
683	0.797	0.954	0.556	0.769
451	0.387	0.974	0.936	0.766
283	0.851	0.952	0.491	0.765
381	0.488	0.970	0.832	0.763
111	0.599	0.964	0.726	0.763
663	0.745	0.954	0.583	0.761
311	0.543	0.965	0.772	0.760
493	0.768	0.954	0.544	0.755
643	0.798	0.950	0.512	0.753
591	0.463	0.970	0.823	0.752
243	0.858	0.948	0.448	0.751
571	0.620	0.959	0.674	0.751
613	0.825	0.948	0.478	0.750
341	0.486	0.966	0.798	0.750
293	0.824	0.951	0.474	0.750
321	0.425	0.971	0.853	0.750
163	0.790	0.952	0.506	0.749
483	0.721	0.954	0.562	0.746
271	0.642	0.959	0.636	0.746
123	0.772	0.953	0.508	0.744
423	0.684	0.956	0.591	0.744
213	0.842	0.946	0.416	0.735
463	0.656	0.955	0.589	0.733
673	0.897	0.941	0.362	0.733
361	0.376	0.970	0.852	0.733
443	0.721	0.950	0.518	0.730
413	0.755	0.948	0.484	0.729
183	0.752	0.951	0.481	0.728
581	0.369	0.970	0.835	0.725
511	0.433	0.965	0.774	0.724
431	0.225	0.976	0.966	0.722
473	0.847	0.941	0.368	0.719
143	0.760	0.947	0.438	0.715
193	0.720	0.951	0.464	0.712
541	0.367	0.966	0.801	0.711
521	0.296	0.971	0.855	0.708
171	0.510	0.959	0.632	0.700
113	0.740	0.945	0.406	0.697
561	0.241	0.970	0.854	0.688
793	0.521	0.955	0.561	0.679
273	0.783	0.940	0.307	0.677
773	0.654	0.942	0.383	0.660
783	0.444	0.955	0.580	0.660
653	0.272	0.962	0.733	0.656
723	0.385	0.956	0.609	0.650
713	0.498	0.949	0.501	0.649
743	0.444	0.951	0.535	0.643
173	0.672	0.939	0.298	0.636
633	0.109	0.968	0.830	0.636
763	0.341	0.955	0.607	0.634
232	0.965	0.918	−0.037	0.615
751	−0.082	0.974	0.942	0.611
132	0.986	0.916	−0.071	0.610
453	0.095	0.962	0.741	0.599
393	0.248	0.955	0.575	0.593

Table A1. *Cont.*

MODEL NO.	$R^2(J)$	$R^2(T_{f,out})$	$R^2(T_{p,out})$	R^2_{tot}
373	0.433	0.942	0.395	0.590
252	0.978	0.911	−0.154	0.578
433	−0.095	0.968	0.837	0.570
152	0.985	0.910	−0.187	0.569
383	0.142	0.956	0.594	0.564
313	0.216	0.949	0.515	0.560
573	0.311	0.942	0.401	0.551
343	0.142	0.951	0.549	0.547
323	0.062	0.957	0.623	0.547
593	0.100	0.955	0.582	0.546
731	−0.323	0.976	0.968	0.541
363	0.002	0.956	0.622	0.527
262	0.968	0.903	−0.306	0.522
222	0.962	0.904	−0.305	0.520
583	−0.020	0.956	0.600	0.512
513	0.065	0.949	0.521	0.512
282	0.955	0.903	−0.333	0.508
292	0.943	0.902	−0.354	0.497
162	0.922	0.902	−0.334	0.497
242	0.960	0.899	−0.372	0.496
543	−0.020	0.951	0.556	0.496
122	0.913	0.903	−0.333	0.494
523	−0.111	0.957	0.629	0.492
212	0.954	0.897	−0.402	0.483
182	0.902	0.902	−0.360	0.481
142	0.908	0.898	−0.399	0.469
563	−0.178	0.956	0.628	0.469
192	0.884	0.901	−0.381	0.468
112	0.898	0.896	−0.429	0.455
351	−0.570	0.974	0.947	0.451
272	0.929	0.892	−0.505	0.439
692	0.522	0.910	−0.186	0.415
753	−0.492	0.963	0.761	0.411
172	0.860	0.891	−0.529	0.407
682	0.461	0.910	−0.162	0.403
622	0.414	0.912	−0.131	0.399
672	0.633	0.898	−0.353	0.393
612	0.505	0.904	−0.238	0.390
642	0.461	0.906	−0.203	0.388
662	0.381	0.911	−0.130	0.387
492	0.398	0.910	−0.169	0.380
551	−0.827	0.974	0.949	0.366
482	0.325	0.911	−0.145	0.364
472	0.529	0.898	−0.338	0.363
422	0.271	0.913	−0.113	0.357
412	0.377	0.905	−0.221	0.353
331	−0.887	0.976	0.969	0.353
733	−0.766	0.969	0.855	0.352
442	0.326	0.907	−0.186	0.349
462	0.231	0.912	−0.112	0.343
772	0.176	0.900	−0.297	0.260
792	−0.019	0.912	−0.123	0.257
531	−1.183	0.977	0.969	0.254
652	−0.222	0.920	0.050	0.249
632	−0.411	0.927	0.177	0.231
782	−0.126	0.913	−0.099	0.230
712	−0.050	0.907	−0.177	0.227
353	−1.087	0.963	0.777	0.218

Table A1. *Cont.*

MODEL NO.	$R^2(J)$	$R^2(T_{f,out})$	$R^2(T_{p,out})$	R^2_{tot}
742	−0.125	0.909	−0.141	0.214
722	−0.206	0.915	−0.066	0.214
762	−0.263	0.914	−0.065	0.195
452	−0.465	0.921	0.071	0.175
372	−0.186	0.901	−0.264	0.150
432	−0.684	0.928	0.198	0.147
333	−1.444	0.969	0.869	0.131
392	−0.444	0.914	−0.087	0.128
553	−1.396	0.964	0.784	0.117
312	−0.483	0.908	−0.141	0.095
572	−0.375	0.902	−0.249	0.092
382	−0.583	0.915	−0.062	0.090
342	−0.580	0.910	−0.105	0.075
322	−0.687	0.916	−0.029	0.067
592	−0.666	0.915	−0.071	0.059
362	−0.761	0.915	−0.027	0.042
512	−0.708	0.909	−0.125	0.025
582	−0.820	0.916	−0.045	0.017
533	−1.796	0.970	0.875	0.016
542	−0.817	0.911	−0.088	0.002
522	−0.937	0.917	−0.011	−0.010
562	−1.019	0.916	−0.010	−0.037
752	−1.250	0.923	0.125	−0.067
732	−1.560	0.931	0.254	−0.125
352	−2.023	0.925	0.169	−0.309
332	−2.422	0.933	0.300	−0.397
552	−2.419	0.926	0.189	−0.435
532	−2.866	0.934	0.320	−0.537

References

1.	Alklaibi, A.M.; Lior, N. Membrane-Distillation Desalination: Status and Potential. *Desalination* **2005**, *171*, 111–131. [CrossRef]
2.	Alkhudhiri, A.; Hilal, N. Membrane Distillation—Principles, Applications, Configurations, Design, and Implementation. In *Emerging Technologies for Sustainable Desalination Handbook*; Elsevier Inc.: Amsterdam, The Netherlands, 2018; pp. 55–106.
3.	Khayet, M. Membranes and Theoretical Modeling of Membrane Distillation: A Review. *Adv. Colloid Interface Sci.* **2011**, *164*, 56–88. [CrossRef]
4.	Wu, Y.; Kong, Y.; Liu, J.; Zhang, J.; Xu, J. An Experimental Study on Membrane Distillation-Crystallization for Treating Waste Water in Taurine Production. *Desalination* **1991**, *80*, 235–242. [CrossRef]
5.	Edwie, F.; Chung, T.-S. Development of Simultaneous Membrane Distillation–Crystallization (SMDC) Technology for Treatment of Saturated Brine. *Chem. Eng. Sci.* **2013**, *98*, 160–172. [CrossRef]
6.	Ali, A.; Quist-Jensen, C.A.; Jørgensen, M.K.; Siekierka, A.; Christensen, M.L.; Bryjak, M.; Hélix-Nielsen, C.; Drioli, E. A Review of Membrane Crystallization, Forward Osmosis and Membrane Capacitive Deionization for Liquid Mining. *Resour. Conserv. Recycl.* **2021**, *168*, 105273. [CrossRef]
7.	Bouchrit, R.; Boubakri, A.; Mosbahi, T.; Hafiane, A.; Bouguecha, S.A.T. Membrane Crystallization for Mineral Recovery from Saline Solution: Study Case Na$_2$SO$_4$ Crystals. *Desalination* **2017**, *412*, 1–12. [CrossRef]
8.	Simoni, G.; Kirkebæk, B.S.; Quist-Jensen, C.A.; Christensen, M.L.; Ali, A. A Comparison of Vacuum and Direct Contact Membrane Distillation for Phosphorus and Ammonia Recovery from Wastewater. *J. Water Process Eng.* **2021**, *44*, 102350. [CrossRef]
9.	Khayet, M.; Matsuura, T. *Membrane Distillation: Principles and Applications*; Elsevier: Amsterdam, The Netherlands; Boston, MA, USA, 2011.
10.	Drioli, E.; Ali, A.; Macedonio, F. Membrane Distillation: Recent Developments and Perspectives. *Desalination* **2015**, *356*, 56–84. [CrossRef]
11.	Laganà, F.; Barbieri, G.; Drioli, E. Direct Contact Membrane Distillation: Modelling and Concentration Experiments. *J. Memb. Sci.* **2000**, *166*, 1–11. [CrossRef]
12.	Mart, L.; Rodr, J.M. On Transport Resistances in Direct Contact Membrane Distillation. *J. Memb. Sci.* **2007**, *295*, 28–39. [CrossRef]
13.	Qtaishat, M.; Matsuura, T.; Kruczek, B.; Khayet, M. Heat and Mass Transfer Analysis in Direct Contact Membrane Distillation. *Desalination* **2008**, *219*, 272–292. [CrossRef]
14.	Thomas, L.C. *Heat Transfer*; Prentice-Hall: Englewood Cliffs, NJ, USA, 1992.

15. Curcino, I.V.; Júnior, P.R.S.C.; Gómez, A.O.C.; Chenche, L.E.P.; Lima, J.A.; Naveira-Cotta, C.P.; Cotta, R.M. Analysis of Effective Thermal Conductivity and Tortuosity Modeling in Membrane Distillation Simulation. *Micro Nano Eng.* **2022**, *17*, 100165. [CrossRef]

16. Cheng, L.; Wu, P.; Chen, J. Modeling and Optimization of Hollow Fiber DCMD Module for Desalination. *J. Memb. Sci.* **2008**, *318*, 154–166. [CrossRef]

17. Olatunji, S.O.; Camacho, L.M. Heat and Mass Transport in Modeling Membrane Distillation Configurations: A Review. *Front. Energy Res.* **2018**, *6*, 130. [CrossRef]

18. Lin, S.; Yip, N.Y.; Elimelech, M. Direct Contact Membrane Distillation with Heat Recovery : Thermodynamic Insights from Module Scale Modeling. *J. Memb. Sci.* **2014**, *453*, 498–515. [CrossRef]

19. Tewodros, B.N.; Yang, D.R.; Park, K. Design Parameters of a Direct Contact Membrane Distillation and a Case Study of Its Applicability to Low-Grade Waste Energy. *Membranes* **2022**, *12*, 1279. [CrossRef]

20. Ali, A.; Quist-Jensen, C.A.; Macedonio, F.; Drioli, E. On Designing of Membrane Thickness and Thermal Conductivity for Large Scale Membrane Distillation Modules. *J. Membr. Sci. Res.* **2016**, *2*, 179–185.

21. Hitsov, I.; Eykens, L.; de Sitter, K.; Dotremont, C.; Pinoy, L.; Van der Bruggen, B. Calibration and analysis of a direct contact membrane distillation model using Monte Carlo filtering. *J. Memb. Sci.* **2016**, *515*, 63–78. [CrossRef]

22. Nagaraj, N.; Patil, G.; Babu, B.R.; Hebbar, U.H.; Raghavarao, K.S.M.S.; Nene, S. Mass Transfer in Osmotic Membrane Distillation. *J. Memb. Sci.* **2006**, *268*, 48–56. [CrossRef]

23. Gryta, M.; Tomaszewska, M. Heat Transport in the Membrane Distillation Process. *J. Memb. Sci.* **1998**, *144*, 211–222. [CrossRef]

24. Alklaibi, A.M.; Lior, N. Heat and Mass Transfer Resistance Analysis of Membrane Distillation. *J. Memb. Sci.* **2006**, *282*, 362–369. [CrossRef]

25. Gryta, M.; Tomaszewska, M.; Morawski, A.W. Membrane Distillation with Laminar Flow. *Sep. Purif. Technol.* **1997**, *5866*, 2–6. [CrossRef]

26. Phattaranawik, J.; Jiraratananon, R.; Fane, A.G. Heat Transport and Membrane Distillation Coefficients in Direct Contact Membrane Distillation. *J. Memb. Sci.* **2003**, *212*, 177–193. [CrossRef]

27. Kim, W.J.; Campanella, O.; Heldman, D.R. Predicting the Performance of Direct Contact Membrane Distillation (DCMD): Mathematical Determination of Appropriate Tortuosity Based on Porosity. *J. Food Eng.* **2021**, *294*, 110400. [CrossRef]

28. Yu, H.; Yang, X.; Wang, R.; Fane, A.G. Analysis of Heat and Mass Transfer by CFD for Performance Enhancement in Direct Contact Membrane Distillation. *J. Memb. Sci.* **2012**, *405–406*, 38–47. [CrossRef]

29. Imdakm, A.O.; Matsuura, T. A Monte Carlo Simulation Model for Membrane Distillation Processes: Direct Contact (MD). *J. Memb. Sci.* **2004**, *237*, 51–59. [CrossRef]

30. Tsai, J.H.; Quist-Jensen, C.; Ali, A. Multipass Hollow Fiber Membrane Modules for Membrane Distillation. *Desalination* **2023**, *548*, 116239. [CrossRef]

31. Quist-jensen, C.A.; Ali, A.; Mondal, S.; Macedonio, F.; Drioli, E. A Study of Membrane Distillation and Crystallization for Lithium Recovery from High-Concentrated Aqueous Solutions. *J. Memb. Sci.* **2016**, *505*, 167–173. [CrossRef]

32. Ali, A.; Quist-Jensen, C.A.; Macedonio, F.; Drioli, E. Application of Membrane Crystallization for Minerals' Recovery from Produced Water. *Membranes* **2015**, *5*, 772. [CrossRef]

33. Ali, A.; Quist-Jensen, C.A.A.; Macedonio, F.; Drioli, E. Optimization of Module Length for Continuous Direct Contact Membrane Distillation Process. *Chem. Eng. Process. Process Intensif.* **2016**, *110*, 188–200. [CrossRef]

34. Bejan, A.; Kraus, A.D. *Heat Transfer Handbook*; John Wiley & Sons: Hoboken, NJ, USA, 2003; ISBN 0471390151.

35. Schofield, R.W.; Fane, A.G. Heat and Mass Transfer in Membrane Distillation. *J. Memb. Sci.* **1987**, *33*, 299–313. [CrossRef]

36. Ali, A.; Tsai, J.-H.; Tung, K.-L.; Drioli, E.; Macedonio, F. Designing and Optimization of Continuous Direct Contact Membrane Distillation Process. *Desalination* **2018**, *426*, 97–107. [CrossRef]

37. Banat, F.A.; Simandl, J. Desalination by Membrane Distillation: A Parametric Study. *Sep. Sci. Technol.* **1998**, *33*, 201–226. [CrossRef]

Localized Heating to Improve the Thermal Efficiency of Membrane Distillation Systems

Alessandra Criscuoli * and Maria Concetta Carnevale

Institute on Membrane Technology (CNR-ITM), Via P. Bucci 17/C, 87036 Rende, CS, Italy
* Correspondence: a.criscuoli@itm.cnr.it; Tel.: +39-0984-492118

Abstract: Membrane distillation (MD) is a thermal-based membrane operation with high potential for the treatment of aqueous streams. However, its implementation is limited and only few examples of MD pilots can be found in desalination. One of the reasons behind this is that MD requires thermal energy for promoting the evaporation of water, which implies higher energy consumption with respect to pressure-driven membrane operations, like reverse osmosis (RO). Recently, among the different methods investigated to improve the thermal efficiency of MD, attempts for obtaining a localized heating of the feed, close to the membrane surface, were carried out. This work reviews experimental activities on the topic, dealing with both modified membranes, used under solar irradiation or coupled to an electric source, and specifically designed heated modules. The main results are reported and points of action for further optimization are identified. In particular, although at an early stage, this type of approach led to improvements in membrane flux and to a reduction of energy consumption with respect to conventional MD. Nevertheless, long tests to ensure a stable performance time, the optimization of operating conditions, the development of methods to control fouling issues, and the identification of the best module design, together with the scale-up of membranes/modules developed, represent the main research efforts needed for future implementation of localized heating strategy.

Keywords: electrical heating; irradiation heating; flux; energy consumption

Citation: Criscuoli, A.; Carnevale, M.C. Localized Heating to Improve the Thermal Efficiency of Membrane Distillation Systems. *Energies* **2022**, *15*, 5990. https://doi.org/10.3390/en15165990

Academic Editor: Chanwoo Park

Received: 26 July 2022
Accepted: 17 August 2022
Published: 18 August 2022

Publisher's Note: MDPI stays neutral with regard to jurisdictional claims in published maps and institutional affiliations.

1. Introduction

Water scarcity today affects many countries worldwide. Climate change, due to the greenhouse effect, significantly increased the number of regions suffering from drought. As an example, as this article is being written, Italy is experiencing one of the driest seasons with the water flows of the most important rivers reduced by 90% in some areas. In addition to the problem of drought, population growth, as well as the intensive industrial activities contributed in recent years to the depletion of fresh water sources available on the planet. The recovery of purified water, to be re-used from polluted streams is thus becoming an urgent need. In this respect, membrane distillation is among the membrane operations most investigated because of the possibility of rejecting all non-volatile species present in the same unit into the feed, leading to a high-concentrated retentate and to a high-quality water (distilled water) as permeate [1–3]. It has been successfully applied in different fields, like ultrapure water production, purification of textile effluents, olive mill waste waters and heavy metal-contaminated waters, juice concentration, brackish and seawater desalination, and brine treatment. The basic principle is to use a hydrophobic microporous membrane for aqueous feed evaporation. In particular, one side of the membrane is in direct contact with the aqueous stream and by creating a difference of vapor pressure at the two membrane sides the liquid starts to evaporate at the feed-membrane pore mouths. Then the formed water vapor migrates through the micropores to be recovered as liquid at the permeate side. The driving force being a difference of vapor pressures rather than a difference of pressures, MD is less affected by osmotic limitations, encountered in RO therefore, higher water recovery factors can be obtained. However, to make the liquid water evaporate there is a need to heat the feed stream. Although MD works at lower temperatures than conventional distillation columns (typical MD temperatures range from 40 to 80 °C), the thermal energy supply represents one of the main obstacles for the implementation of MD at a commercial scale. The energy consumption per produced permeate can be reduced by acting both on the productivity of the process and on the effective use of the thermal energy supplied. Given that MD is a thermal-driven operation the trans-membrane flux and thus, the productivity (defined as the amount of produced permeate) depends not only on the membrane properties but also on the thermal efficiency (e.g., temperature established at the membrane surface for water evaporation). Therefore, an improvement of thermal energy use during the operation is crucial. In MD thermal losses into the environment through the pipelines and the membrane module can occur. Furthermore, the permeate acquires the heat of condensation of the water vapor. If not recovered, this is another step where the heat supplied is lost. The thermal losses to the outside can be decreased by a proper insulation of the circuit and of the module, and choosing low-conductive materials will also assist in their realization. Concerning the heat recovery from the permeate, the use of a heat exchanger in which the feed is pre-heated while the permeate is cooled, as well as the design of modules with internal heat recovery are possible solutions [4–10]. In addition to the reduction of heat losses, the energy efficiency of MD is also based on the minimization, especially at the feed side, of the temperature polarization which consists of the temperature gradient created between the bulk of the stream and the membrane surface. For an optimal evaporation process, it is desirable to have the membrane surface at the same temperature as the bulk, so as to effectively use the warm stream. However, the membrane temperature is often lower than that of the bulk, due to the heat transfer resistance offered by the boundary layer (Figure 1), therefore, the water evaporation is lower than that achievable at the feed bulk temperature. Moreover, during the MD process, the temperature at the membrane surface further decreases because of the evaporative cooling of the water in contact with the membrane.

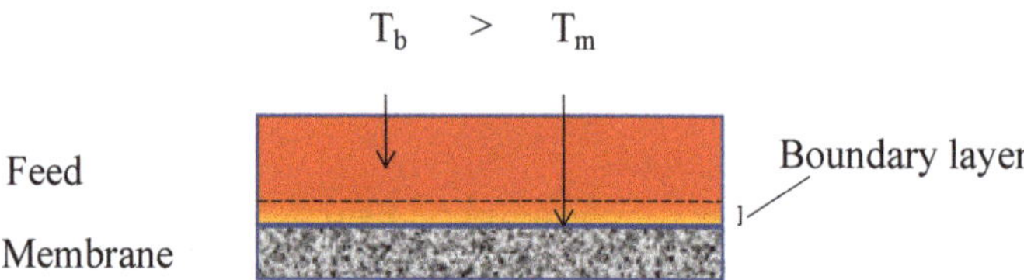

Figure 1. Temperature difference at the feed side, due to the boundary layer resistance.

An increase of the feed flow rate and/or the realization of modules with baffles/turbulence promoters could enhance the turbulence inside the module, thus reducing the boundary layer resistance [11–14]. However, pressure drops must be carefully controlled, in order to not overcome the liquid entry pressure value of the membrane with a consequent wetting of micropores. On the other hand, the temperature reduction caused by the water evaporation could not be avoided by the above strategies. Recently, the possibility of applying localized heating was investigated as a new approach to enhance the thermal performance of MD through the reduction of the temperature polarization at the feed side. In particular, both the heating inside the module and the direct heating of the membrane surface were studied. In the first case, heat was supplied to the module, without changing the membrane properties, while in the second case the membrane itself was also modified. Both approaches had the aim of increasing the temperature of the feed at the membrane surface, thus enhancing the water evaporation and the permeate production. By acting on the temperature close to the membrane surface, the cooling effect of the evaporation can also be better compensated. Solar and electrical energy sources were used to provide heat. Research in the field has significantly increased in last years and in this contribution, the main experimental activities carried out on localized heating are presented and discussed. Future research needs are also underlined.

2. Basic Indicators in Membrane Distillation

The performance of membrane distillation can be evaluated in terms of some specific indicators (well-known and conventionally used in MD), like the water vapor transmembrane flux, which is linked to the process productivity, the temperature polarization factor, which takes into account the difference between the feed temperature in the bulk liquid and at the membrane surface, as well as the indicators linked to the energy consumption of the process and to the thermal energy use for evaporation. When solar irradiation is employed for localized heating another indicator, solar efficiency, is introduced. In the following section, definitions and main equations to be used are reported.

Permeate flux (J)

The distillate flux J is calculated using the following equation:

$$J = \frac{m}{A_{mb} \cdot t} \tag{1}$$

where:

J = distillate flux (L·m^{-2} h^{-1}) or (kg·m^{-2} h^{-1});
m = mass of distillate produced (kg);
A_{mb} = membrane area (m^2);
t = time (h).

High permeate fluxes are desired to ensure high productivities (high permeate production) of the system.

Temperature Polarization Factor (TPF)

The *TPF* (temperature polarization factor) is defined as:

$$TPF = \frac{T_f^m}{T_f^b} \times 100 \tag{2}$$

where:

T_f^m = membrane surface temperature at the feed side (°C);
T_f^b = bulk temperature of the feed (°C).

TPF < 1 indicates that the temperature at the membrane surface is lower than that of the feed bulk. This implies that the evaporation occurs at a lower temperature, with a consequent lower permeate flux.

Specific Energy Consumption (SEC)

SEC is defined as the amount of total energy supplied (thermal and electrical) with respect to the produced distillate (kW·kg^{-1}). In a formula:

$$SEC = \frac{Q_T}{m} \tag{3}$$

where:

Q_T = total energy supplied (kW);
m = mass of distillate produced (kg).

Low *SEC* values are desired to work with low energy consumptions and high productivities of the system.

Gain Output Ratio (GOR)

GOR is a key performance indicator which shows the relation between the energy needed for the feed evaporation and the overall thermal energy supplied. In a formula:

$$GOR = \frac{m_d \cdot \Delta H}{Q_H \cdot}$$

(4)

where:

m_d = distillate flow rate (kg·h^{-1});
ΔH = enthalpy of vaporization (kJ·kg^{-1});
Q_H = overall thermal energy supplied (kJ·h^{-1}).

A $GOR > 1$ indicates a good use of the heat supplied. Its value depends on the membrane and module properties, on the operating conditions, and on the heat recovery systems adopted. At lab scale a $GOR < 1$ is often registered [15], while at a larger scale a value > 10 can be obtained.

Solar Efficiency

The solar efficiency of the process is defined as the ratio between the energy used for water evaporation and the overall solar irradiance. In formula [16]:

$$Solar\ Efficiency = \frac{J \cdot \Delta H}{I}$$

(5)

where:

J = distillate flux (kg·m^{-2}·s^{-1});
ΔH = enthalpy of vaporization (kJ·kg^{-1});
I = incident light intensity (kJ·m^{-2}·s^{-1}).

High solar efficiencies are desired to effectively use the solar irradiance supplied to the system.

3. Localized Heating with Modified Membranes

Membranes were modified in order to make themselves "heating units" inside the modules. Modifications were carried out on the membrane surface only, as well as on the whole membrane structure. Depending on the type of membranes produced, both solar and electrical energies were supplied, so as to warm up the membrane and to enhance the water evaporation (Figure 2). In the following part, the main information on the type of membranes prepared, experimental tests, and improvements obtained with respect to traditional MD are reported. The most relevant results are summarized in Tables 1 and 2.

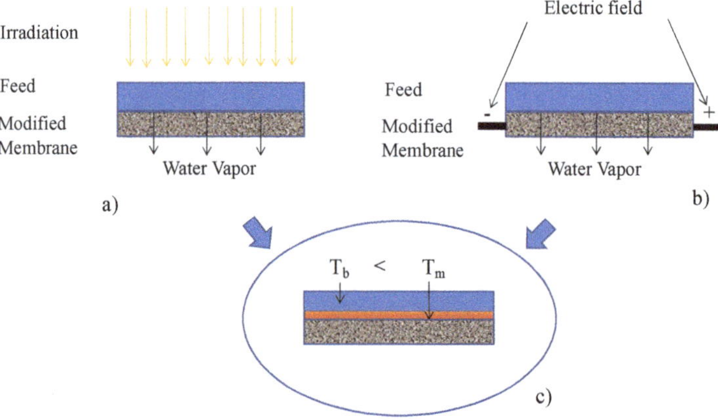

Figure 2. Modified membranes under (**a**) irradiation, (**b**) electrical field, and (**c**) temperature difference at the feed side.

3.1. Electrical Heating on Modified Membranes

Song et al. [17] employed a nichrome resistance wire (NRW) inside a polyvinylidene fluoride (PVDF) hollow fiber membrane (HFM) to realize an electro-thermal PVDF/NRW HFM module with an effective area of 20.17 cm^2. The experimental tests were carried out using a 3.5 wt% NaCl aqueous solution as feed and the vacuum membrane distillation (VMD) configuration (vacuum pressure: -0.04 MPa). The feed was heated up to 70 °C by a hot water bath. When a low direct current of 0.15 A was applied, a 2.5-fold increase of the permeate flux was measured with respect to that obtained without the application of electrical current.

A high-quality hexagonal boron nitride (hBN) nano coating was coated on stainless-steel wire cloth (SSWC), and the obtained hBN-SSWC was successively attached to a commercial PVDF membrane by Zuo et al. [18], to create a Joule heater in MD (SHMD surface heating membrane distillation). The flat SHMD module had a membrane area of 1.6 cm^2 and by starting with a 100 g/L NaCl feed a high-concentrated brine of 302.9 g/L was obtained, corresponding to a 67% single pass-recovery with a percentage of input energy utilized to produce vapor of 57%, when the energy input was 50 kW·m^{-2}. The long-term stability of the membrane was also assessed for 100 h of operation and a spiral-wound SHMD module was produced with a percentage of input energy utilized to produce vapor of 79.1% and 875.8 kW·m^{-3} of energy consumption without heat recovery.

Dudchenko et al. [19] realized a stable dual-layer structure with hydrophilic–hydrophobic characteristics through the deposit of CNT (carbon nanotube)–PVA (polyvinyl alcohol) films onto a hydrophobic polytetrafluoroethylene (PTFE) membrane (area of 450 cm^2). By supplying a current, the high temperature reached at the membrane surface allowed single-pass recovery up to 100% during MD desalination tests to be registered. A GOR of 0.55 was obtained and no degradation of the CNT deposit was observed.

To cover the situations in which the solar intensity is not at the desired value (e.g., cloudy days), photothermal and Joule heating MD were coupled by using a composite membrane made of three layers: a bottom PVDF layer, a middle multi-walled carbon nanotube (MWCNT) layer, and a top polydimethylsiloxane (PDMS) layer [20]. The module's upper part had a transparent optical glass window to allow irradiation by sunlight, while the electricity was supplied through an alternating-current power source. In the bottom part, a condensation chamber was used to condense the vapor. The combined inputs of solar and electric energy allowed it to work under a constant total power. Interestingly, at the same input-power density, the Joule heating MD led to higher feed-water temperatures than the solar energy with consequent higher trans-membrane fluxes. To further optimize the performance of the Joule heating MD, Huang et al. [21] investigated different strategies. First of all, the feed was not recirculated but kept in the upper part of the cell, so as to heat just the water close to the membrane rather than the whole stream. By acting on the power supply and on the conductive layer area it was possible to further increase the temperature for the evaporation. Moreover, a three-level heat recovery design, where the condensing heat of the upper level was recovered by the feed stream of the lower level, was studied. In this case, only the first level contained the PDMS/MWCNT/PVDF membrane and underwent electrical energy supply, while the second and the third levels were only equipped with PVDF membranes. By moving from a single-level to a three-level system, the water flux increased by 2-fold with a GOR of 1.89.

Ahmed et al. [22] prepared an electrically conductive carbon nanostructure (CNS) coated polypropylene (PP) membrane and tested it in a direct contact membrane distillation (DCMD) configuration (membrane area, 22.5 cm^2) with and without applying current, while the circulating feed was externally heated. At 60 °C, the electrical current supply led to a 61% increase of flux (22.9 vs. 14.2 kg·m^{-2} h^{-1}) and to a reduction of the specific energy consumption higher than 50% (1.7 vs. 3.7 kW·m^{-3}).

In the study of Li et al. [23] a new methodology that used a composite RGO (reduced graphene oxide)-PTFE (polytetrafluoroethylene) membrane for reverse Joule heating air gap MD (AGMD) was presented. In particular, the layer of RGO Joule heating was located at the air gap side and the contact with the feed brine, which could cause water splitting and RGO degradation in saline environments, was prevented by the PTFE membrane. A membrane module of 29.65 cm^2 was used to carry out the experiments and it was demonstrated that the permeate flux was kept stable during 115 h of testing at a value around 1 (as normalized flux), indicating that the RGO layer had good stability.

Subrahmanya et al. [24] tested a graphene-PVDF flat membrane Joule heater for VMD desalination. The graphene and PVDF content were varied from 10 to 25% and from 1 to 10%, respectively, with respect to the solvent N-Methyl-2-pyrrolidone (NMP). Desalination tests were carried out with a module of 778.54 mm^2 membrane area and best results were obtained for the graphene 2.5-PVDF1 membrane, working at the lowest feed flowrate (1 mL/min) and with an energy supply of 2 W, for which a temperature of 56 °C was registered on the membrane surface and the GOR was 5.72.

Metallic (stainless steel—SS) hollow fiber membranes were coated or impregnated with PDMS and used in sweep gas membrane distillation (SGMD) for water evaporation tests under electrical heating [25]. The membrane area at the lumen side, where dry air flowed, was 140 cm^2 and experiments were carried out at different water inlet temperatures. In all tests, by supplying electrical energy, water evaporation flux enhancement factors were registered, ranging from 1.1 to 1.4.

Anvari et al. [26] have studied a new IH (induction heating)-VMD system based on a 'self-heating' composite membrane realized by spraying a coating of iron oxide-carbon nanotubes on a hydrophobic PTFE commercial membrane. The induction heating allowed the membrane surface to heat in a contactless mode. The membrane module (32 cm^2) was made of nylon and was used for both IH-VMD and traditional VMD tests (with a commercial unmodified PTFE membrane) in the same operating conditions. For a 35 g/L NaCl feed, the IH-VMD system led to an 8-fold higher trans-membrane flux (4 vs. 0.5 kg·m^{-2} h^{-1}) and to a 6-fold lower specific energy consumption (197 vs. 1202 W·kg^{-1}). The GOR was 3.45.

3.2. Irradiation Heating on Modified Membranes

The studies of Politano et al. [27,28] illustrated a plasmonic photothermal MD process that employed asymmetric PVDF flat-sheet microporous membranes in which metallic Ag NPs (nanoparticles) were incorporated in variable percentages (from zero to 25%). The authors carried out tests using the VMD configuration with a UV lamp (with a wavelength of 366 nm) to irradiate 21.24 cm^2 of membrane area through a quartz window. The best results were obtained with a 25% Ag NPs load: for a 0.5 M feed solution at an initial temperature of 303 K, the bulk temperature was increased by about 4 K, the trans-membrane flux (25.7 kg·m^{-2} h^{-1}) was 9-fold higher than the corresponding value for the unloaded membrane, while the temperature polarization factor (TPF) moved from 98.25% (unloaded membrane) to 106.5%, due to the interface temperature being higher than that in the bulk.

Ag photothermal nanoparticles (Ag NPs) were also incorporated in different amounts into a hydrophobic PVDF nanofibrous membrane by Ye et al. [16] to carry out tests of ultraviolet light driven DCMD. The electrospinning technique was applied and an effective membrane area of 12 cm^2 was located inside a module with a quartz window. The best performance was reached with the membrane containing 20 wt% Ag NPs, leading to the highest flux and solar efficiency (53%) during 60 h of testing.

A mixed matrix of hydrophobic photoactive membrane was prepared by Pagliero et al. [29] by dispersing carbon black (CB) in a PVDF dope solution and using NIPS (non-solvent induced phase separation) as a preparation technique. Then, VMD tests were carried out on a 30 cm^2 membrane area located in a membrane module made of PMMA (transparent polymethylmethacrylate) which was irradiated. The best performances were obtained with the membrane containing 7.5 wt% CB, with a 2-fold increase of the trans-membrane flux with respect to the pristine PVDF membrane (2.3 vs. 1 kg·m^{-2} h^{-1}).

Dongare et al. [30] used CB NPs for the preparation of a photothermal membrane consisting of two layers: a hydrophilic polyvinyl alcohol (PVA) coating, with a thickness of 25 μm, deposited onto a commercial PVDF membrane. Tests were carried in a DCMD configuration (countercurrent mode) sending at the feed side a 1% NaCl solution, whereas at the distillate side flowed deionized water. A quartz window (3.3 cm × 6.8 cm) allowed the irradiation of the membrane at the feed side without additional heat sources, leading to a solar efficiency of over 20%. When compared with the traditional MD, at parity of operating conditions, a higher flux was obtained at low feed velocity (0.3 vs. 0.05 kg· m^{-2} h^{-1}).

Said et al. [31] carried out tests with a membrane prepared by directly coating functionalized CB NPs on a commercial hydrophobic PTFE membrane. The membrane module had an area of 0.17 m^2 and was equipped with a transparent window of durable Plexiglas. A polycrystalline solar panel was used to supply energy and in cloudy conditions a reduction of the experimental flux was registered (0.12 kg·m^{-2} h^{-1} at 88 W·m^{-2}) with an average value of 0.55 kg·m^{-2} h^{-1}.

Wu et al. [32] used PVDF membranes coated with CB NPs or SiO$_2$/Au for direct solar DCMD tests on a 28.3 cm^2 membrane area, which was irradiated through a quartz window with simulated sunlight obtained using six halogen tungsten lamps. Ultrapure water was sent at the cold side and 1% NaCl solution at 35 °C was used as the feed. The sample with a CB coating density around 0.14 g m^{-2} led to the highest permeate flux with a 15% increase with respect to that achieved in tests without irradiation.

A CB NPs-coated PVDF membrane (25 cm^2) was developed and used by Tanvir et al. [33] in a passive, single-stage, permeate-side-heated solar MD unit. In this case, the feed filled a bottom chamber and was in contact with the PVDF side of the membrane while the CB coating was irradiated. The evaporated water was condensed at the top side in a condensing chamber equipped with a reflective cover. Seawater, canal water, and wastewater were treated, and initial fluxes were 1.48, 1.34, and 1.32 kg·m^{-2} h^{-1}, respectively. Corresponding operating time were 32, 18, and 10 days, after which, wetting was observed due to scaling, organic fouling, and the presence of surfactants, respectively. Interestingly, the produced permeates were of high-quality, but contained dichloromethane and methyl ethyl ketone, probably originating from the acrylic cement used for fixing the system. Under natural sunlight (652 W·m^{-2}) the energy efficiency was 67.5%.

In the work of Chen et al. [34] a PVDF membrane surface was coated by 1H,1H,2H,2H-Per-fluorodecyltriethoxysilane (FAS17) modified CB NPs, in order to combine the photothermal activity of CB NPs with omniphobic properties conferred by the presence of the fluorinated species. DCMD desalination tests were carried out on a flat membrane (31 cm^2), which was located in a module and was irradiated through a quartz window. The feed was sent to the module at 35 °C. Under irradiation, the composite membrane led to a 25% increase in flux with respect to the pristine PVDF (3.19 vs. 2.56 kg·m^{-2} h^{-1}). Moreover, when compared with traditional DCMD (without irradiation) a 55.6% increase of the input energy used to produce fresh water was registered. Tests with a model surfactant solution of sodium dodecyl sulfate (SDS) confirmed the omniphobic character of the produced membrane. The membrane was not affected by wetting in all the concentration ranges investigated (0.1–0.4 mM), while the PVDF membrane started to be wetted by the 0.2 mM SDS feed. This result is quite interesting for the treatment of wastewaters containing detergents, soaps, and surfactants, for which typical MD membranes suffer from loss of hydrophobicity.

Gong et al. [35] worked with a multilevel-roughness membrane by immobilizing a nanoparticle-assembled superstructure on a nanofibrous membrane. The particular membrane structure was obtained spraying a FTCS (fluorododecyltrichlorosilane)-CB (carbon black) suspensions at different concentration values (the optimal percentage was 2.0%) on a PVDF membrane. The DCMD configuration was used to carry out the experimental tests with natural seawater and oil-contaminated solutions as feed, and illumination was provided though a quartz window. During 48 h of testing, membrane stability and antifouling behavior were assessed, with solar efficiencies ranging from 55 to 67% for illuminations of 1 kW·m^{-2} and 10 kW·m^{-2}, respectively.

In the study of Wu et al. [36] a polydopamine (PDA)—coated PVDF membrane was employed in a solar-driven membrane distillation process. The PDA-PVDF membrane was subjected to a treatment of fluoro-silanization (FTCS-PDA-PVDF membrane), to increase its hydrophobicity, and was used in DCMD tests on a 0.5 M NaCl solution. Under 0.75 kW·m^{-2} and 7.0 kW·m^{-2} irradiation, the FTCS-PDA-PVDF membrane led to a corresponding trans-membrane flux 5-fold (0.49 vs. 0.09 kg·m^{-2} h^{-1}) and 19-fold higher (4.23 vs. 0.22 kg·m^{-2} h^{-1}) than that of the FTCS-PVDF membrane. Moreover, a solar efficiency of 45% and 41% were registered, under 0.75 kW·m^{-2} and 7.0 kW·m^{-2} irradiation, respectively.

Ghim et al. [37] developed a membrane by using the spray-coating with deposition of graphene onto a hydrophobic PTFE membrane with polymerized dopamine (PDA) and trichloro (1H,1H,2H,2Hperfluorooctyl) silane (FTCS). The FTCS-PDA/graphene/PTFE membrane was used in the first (upper) layer of a multi-layer stacked membrane module with air gaps, realized to recover the latent heat of vaporization, while the other layers were equipped with PTFE membranes. During tests, the feed (a salty solution) was stagnant (3 mm thick) to reduce heat losses. With four recovery layers and under 0.75 kW/m^2 of irradiation, a 105% solar efficiency was registered.

Huang et al. [38] used photothermal PVDF/ATO (antimony doped tin oxide) hybrid nanofiber membranes in VMD tests. The feed was sent to the module at 70 °C, using an external heater. When the membrane area of 19.63 cm^2 was irradiated at the highest ATO percentage (5%) a 3-fold increase of flux was registered with respect to that achieved without irradiation, sending a salty solution as feed (27 vs. 8 kg·m^{-2} h^{-1}).

A membrane with photothermal characteristics using TiN (titanium nitride) NPs, capable of absorbing sunlight and converting it into energy, was developed by Zhang et al. [39]. The support was a PVDF flat-sheet membrane, and the technique used to produce the membrane consisted of a facial two-step method: electrospun onto PVDF membrane of a TiN/PVA suspension and crosslinking of this PVA layer. The produced membrane had an area of 19.625 cm^2 and was tested in AGMD on salty solutions. The best results were achieved with the 10 wt% TiN NPs, with a solar efficiency of 64%, and an increase of 65.8% in flux (0.94 vs. 0.57 kg·m^{-2} h^{-1}) with respect to the PVDF membrane. Moreover, the TPC increased from 95.90% (PVDF) to 97.21% and the membrane was stable after 240 h of testing.

The performance of a Fe$_3$O$_4$/PVDF-HFP (co-hexafluoropropylene) membrane with high porosity in solar MD desalination, was studied by Li et al. [40]. The tests were carried out with a membrane area of 37.5 cm^2 under different irradiations for both the composite and the pristine PVDF membrane. At 1 kW·m^{-2} and 3 kW·m^{-2} irradiation values, the composite membrane showed a permeate flux 4 (0.97 vs. 0.26 kg·m^{-2} h^{-1}) and 6 (2.9 vs. 0.48 kg·m^{-2} h^{-1}) times higher than the PVDF-HFP one. The corresponding solar efficiencies were 53% and 59%, respectively. The prepared membrane was stable in 10 days of testing and showed interesting performance also as a pilot scale (at 3 kW·m^{-2}, a flux of about 22 kg·m^{-2} h^{-1} was registered).

Huang et al. [41] prepared a PDMS/CNT/PVDF trilayer membrane. Desalination MD tests were carried out in a two-level device (16 cm^2 membrane area) where the top level contained the trilayer membrane in contact with the feed, which was covered with glass for the irradiation. The second level was equipped with a PVDF membrane and used the heat from the permeate produced in the first level for heating the feed. Both levels had condensation chambers to condense the water vapor. A higher productivity than the pristine PVDF, linked to the photothermal activity of the trilayer membrane, was observed for tests carried out on only one level (the top one). In this case, a 2.4-fold increase in flux was measured (0.37 vs. 0.89 kg·m^{-2} h^{-1}), while the solar efficiency was 24.7% and 59% for the PVDF and the trilayer membrane, respectively.

Han et al. [42] developed, for the first time, a bio-derived membrane to be used in solar driven MD. Eggshell was the starting material from which a carbonized eggshell membrane (cESM) was produced and functionalized with carbon nanotubes (cESM-CNTs). When used under irradiation in DCMD tests on salty solutions at different concentrations, a stable trans-membrane flux was obtained (2.52 times higher than that of the PVDF membrane —1.11 vs. 0.42 kg·m^{-2} h^{-1}) with a solar efficiency greater than 75.6% (vs 31.5% of the PVDF).

Tan et al. [43] used MXene as a coating on a PVDF membrane together with PDMS, to confer both photothermal and anti-fouling characteristics. DCMD tests in counter-current flow mode were carried out on a feed containing 200 mg/L bovine serum albumin (BSA) and 10 g/L NaCl. The membrane area was 37 cm^2 and was placed into a module made of acrylic. Tests were carried out by recirculating the feed at 65 °C. In prolonged tests (21 h), a reduction of 12% of the heat energy input per unit volume distillate and around 60% of the flux decline was registered, with respect to the uncoated membrane.

MXene was coated also onto commercial PTFE membranes by Mustakeem et al. [44] and tested as a self-heating source using only irradiation. The DCMD membrane module was in acrylic, housed 25 cm^2 of membrane area, and was tested on salty solutions at different concentrations. The best performances were obtained at the lowest feed concentration (0.36 g/L), at which a solar efficiency of 65.3% was registered. By increasing the feed salinity, both the vapor flux and the solar efficiency decreased, due to the higher salt amount on the MXene nanosheets which provoked a scattering of the incident light, with a consequent lower temperature at the membrane surface. Already at 10 g/L, the vapor flux decreased by ca. 40%, while the photothermal efficiency was around 38%.

Table 1. Electrical heating on modified membranes (best results).

Heating/Membrane Material	MD Configuration	Feed	Energy Supply	Flux (kg·m^{-2} h^{-1})	SEC (kW·kg^{-1})	Refs.
NRW/PVDF	VMD	3.5 wt% NaCl	3.15 W	14	11.86 [a]	[17]
hBN-SSWC/PVDF	DCMD	100 g/L NaCl	1–50	0.32–42.7	* calc: 3–1.17	[18]
CNT-PVA/PTFE	DCMD	100 g/L NaCl	50 W	7.5	1.25	[19]
PDMS-multiwalled CNT (MWCNT)/PVDF	Distillation with condensation chamber	3.5 wt% NaCl	0.4–1.6 W	0.24–1.1	n.a.	[20]
PDMS-multiwalled CNT (MWCNT)/PVDF	Three-level distillation with condensation chamber	3.5 wt% NaCl	1.6 W	2.77	0.36	[21]
CNS/PP	DCMD	10 g/L NaCl	50.4 W	22.9	1.7 [a]	[22]
RGO/PTFE	AGMD	35 g/L NaCl	5.5 W	1.1	n.a. [a]	[23]
Graphene/PVDF	VMD	3.5 wt% NaCl	2 W	23.44	0.11	[24]
SS-PDMS	SGMD	Water	12 W	0.11	n.a. [a]	[25]
Fe-CNT/PTFE **	VMD	35 g/L NaCl	0.781 kW·m^{-2} (2.46 W)	4	0.2	[26]

* calculated as the ratio between the Energy Supply for electrical localized heating (kW·m^{-2}) and the Flux (kg·m^{-2} h^{-1}). ** Induction heating. [a] external heated feed.

Table 2. Irradiation heating on modified membranes (best results).

Heating/Membrane Material	MD Configuration	Feed	Energy Supply (kW·m^{-2})	Flux (kg·m^{-2} h^{-1})	SEC * (kW·kg^{-1})	Refs.
Ag/PVDF	VMD	0.5 M NaCl	23	25.7	0.90 [a]	[27,28]
Ag/PVDF	DCMD	3.5 wt% NaCl	3.2	2.5	1.28	[16]
CB/PVDF	VMD	Deionized water	0.675	2.3	0.29	[29]
CB-PVA/PVDF	DCMD	1 wt% NaCl	0.7	0.3	2.3	[30]
CB/PTFE	VMD	40 g/L NaCl	0.088–1	0.12–0.77	0.73–1.30	[31]
CB or SiO$_2$-Au/PVDF	DCMD	1 wt% NaCl	1.37	6.12	0.22 [a]	[32]
CB/PVDF	Permeate-side-heated solar MD unit.	Seawater, canal water, wastewater	1.8	1.48, 1.34, 1.32	1.21, 1.34, 1.36	[33]
FAS17-CB/PVDF	DCMD	35 g/L NaCl	1	3.19	0.31 [a]	[34]
FTCS-CB/PVDF	DCMD	seawater	1–10	0.78–9	1.28–1.11	[35]
FTCS-PDA/PVDF	DCMD	0.5 M NaCl	0.75–7	0.49–4.23	1.53–1.65	[36]
FTCS-PDA-graphene/PTFE	AGMD	0.5 M NaCl	0.75	1.17	0.64	[37]
ATO/PVDF	VMD	3.5 wt% NaCl	n.a. (100 W power)	27	n.a. [a]	[38]
TiN-PVA/PVDF	AGMD	35 g/L NaCl	1	0.94	1.06	[39]
Fe$_3$O$_4$/PVDF-HFP	DCMD	3.5 wt% NaCl	1–3	0.97–2.9	1.03 [a]	[40]
PDMS/CNT/PVDF	Two-level distillation with condensation chamber	3.5 wt% NaCl	1	1.43	0.7	[41]
PDMS-multiwalled CNT (MWCNT)/PVDF	Distillation with condensation chamber	3.5 wt% NaCl	0.25–1	0.13–0.92	1.92–1.09	[20]
cESM-CNTs/PVDF	DCMD	2.9–35 g/L NaCl	1	1.11	0.90	[42]
MXene/PVDF	DCMD	200 mg/L BSA in 10 g/L NaCl	5.8	10	0.58 [a]	[43]
MXene/PTFE	DCMD	0.36 g/L NaCl	1	0.77	1.30	[44]

* calculated as the ratio between the Energy Supply for irradiation (kW·m^{-2}) and the Flux (kg·m^{-2} h^{-1}). [a] external heated feed.

3.3. Some Remarks

Localized heating of modified membranes was mainly investigated for the treatment of salty solutions. While fluxes and energy efficiency varied for the different studies, salt was always well rejected, leading to a purified permeate. Flat membranes were mainly developed (often starting from PVDF as pristine membrane material), and small membrane areas were often tested in lab-scale MD experiments. A direct comparison of the main results achieved is not easy, as they depend on various factors, such as the MD configuration, module design and operating conditions (energy supply, external heating of the feed, feed flow, etc.). Nevertheless, all studies confirmed the improvement of flux and the reduction of the energy consumption due to the higher temperature at the membrane surface. The electrical heating of modified membranes flux ranged from 0.11 to 42.7 kg·m^{-2} h^{-1}, with SEC varying from 0.11 kW·kg^{-1} to 11.86 kW·kg^{-1} kg. For the irradiated modified membranes, flux and SEC ranging from 0.12 to 25.7 kg·m^{-2} h^{-1} and from 0.22 to 2.3 kW·kg^{-1}, respectively, were registered. It was demonstrated that, at the same input-power density, the Joule heating MD is able to lead to higher feed-water temperatures than the solar energy, with consequent higher trans-membrane fluxes. The combination of electrical and solar heating can also be an interesting option to cover the periods with low solar radiation (e.g., cloudy periods). When comparing photothermal membranes to conductive ones, it must be noted that it is necessary to act both on the membrane preparation step and on the module design in order to allow the irradiation of the membrane surface. Thus, an extra step is present for their testing in MD applications (see Figure 3).

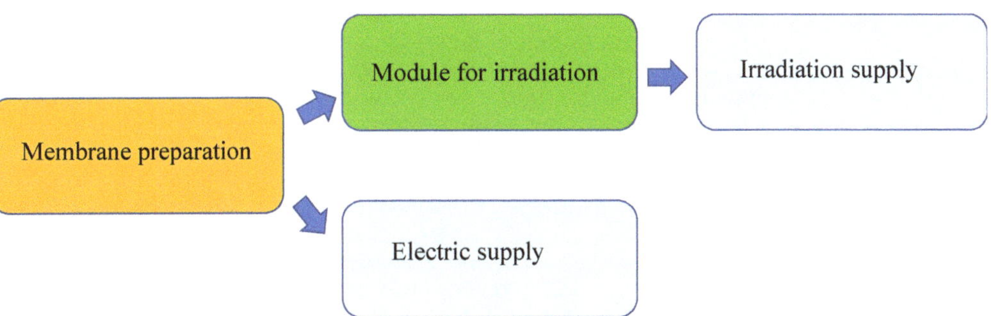

Figure 3. Main steps for localized heating on modified membranes.

4. Localized Heating Inside the Module

Different strategies were developed to heat the feed inside the module-self, through the supply of solar or electrical energy. In this case, commercial membranes were directly used without modification. In the following section, the main information on the type of module developed, experimental tests, achievements, and improvements obtained with respect to traditional MD heating (bulk feed heating) are reported. The most relevant results are summarized in Table 3, while Figure 4 depicts the investigated approaches.

Photothermal nanofluids are known to be able to absorb solar radiation and to transform solar energy into heat. A possible way to improve the solar energy use inside the module is to employ these materials in the feed of a solar powered membrane distillation (SPMD) process. This approach was investigated by Zhang et al. [45], who studied the effect of TiN (titanium nitride) NPs in AGMD tests on salty solutions. To minimize heat losses, the feed was kept static on the membrane surface (PVDF, 19.625 cm^2 membrane area) which was irradiated through a quartz window. The TiN content was varied, and the highest flux was obtained at 100 mg/L TiN. Under 1 kW·m^{-2} irradiation, a 57.4% increase of the solar energy utilization efficiency (50.5% vs. 32.1%), together with a 57% increase of flux (0.74 vs. 0.47 kg·m^{-2} h^{-1}), were registered with respect to the use of a base fluid (salty aqueous solution).

Schwantes et al. [46] proposed a configuration, named the feed gap air gap membrane distillation (FGAGMD), in which the feed was heated inside the module through a polymer film in contact with a heating stream while the permeate side worked with the traditional air gap configuration. The authors developed two plate and frame modules equipped with PTFE membranes (single module membrane area: 8 m^2) to which a salty feed (at various concentrations) was sent. With respect to the traditional spiral wound AGMD module, a 155 g/L NaCl feed in the new system led to a 9% improvement in flux (1.2 vs. 1.1 kg·m^{-2} h^{-1}), a 15-fold higher recovery ratio (45 vs. 3%), and a 9% higher thermal efficiency (50 vs. 46%), defined as the ratio between the energy needed for the feed evaporation and the overall heat transported into the module, while the GOR was lower (1.1 vs. 1.4).

A multi-stage membrane distillation (MSMD) module integrated on the backside of a commercial solar cell, so as to use the waste heat of the solar cell as heat source for MD, was realized by Wang et al. [47]. The module contained hydrophobic electrospun porous polystyrene (PS) membranes with a membrane area of 16 cm^2. Each stage consisted of a top thermal conductive layer (to transfer the heat to the liquid feed), a hydrophilic porous layer (where the evaporation occurs), a hydrophobic porous layer (through which the water vapor is transported), and a condensation layer (where the water vapor condensed). The first stage received heat from the solar panel, while in the others, the liquid water was warmed up by the latent heat of water vapor released during the condensation. With seawater as feed, by working in static conditions, the flux was slightly higher than in the configuration with feed cross-flow, due to a lower heat loss (see Table 3 for values). However, salt accumulation in the device was observed, and washing cycles were needed.

Mustakeem et al. [48] compared the traditional DCMD configuration with three types of localized heating: localized heating cross-flow (LHCF), localized heating dead-end (LHDE) with no feed circulation, and localized heating dead-end with intermittent feed channel flush (LHIF). In all cases, the membrane was in PTFE and the membrane area was 213 cm^2. The localized heating was realized locating a heating coil close to the membrane surface. In the LHCF design, the feed (Red Sea water) was re-circulated, while in the LHDE and LHIF designs, the feed filled the module by gravity. In all cases, higher flux (from 10.2 to 75%) and GOR (from 78 to 150%) and lower specific energy consumption (from 44 to 57%) than the DCMD configuration were obtained, with the LHIF design showing the best efficiency, due to the coupling of the low heat losses to the low fouling of the membrane surface. Specifically, the flux increased from 5.6 kg·m^{-2} h^{-1} (conventional DCMD) to 6 kg·m^{-2} h^{-1} (LHCF), 7.2 kg·m^{-2} h^{-1} (LHDE), and 9.8 kg·m^{-2} h^{-1} (LHIF); the GOR moved from 0.24 (conventional DCMD) to 0.6 (LHIF); the SEC decreased from 2762 kW·m^{-3} (conventional DCMD) to 1183 kW·m^{-3} (LHIF).

Table 3. Localized heating inside the modules (best results).

Heating Type	MD Configuration	Feed	Energy Supply (kW·m^{-2})	Flux (kg·m^{-2} h^{-1})	SEC (kW·kg^{-1})	Refs.
Photothermal nanofluid (TiN)	AGMD	35 g/L NaCl	1–5	0.74–2.77	* 1.35–1.8	[45]
Heating with a heating solution through a polymer film	FGAGMD	155 g/L NaCl	n.a.	1.2	n.a.	[46]
Solar cell-Photovoltaic panel	Three-stage MD	3.5 wt% NaCl	1	1.71 (dead-end)-1.65 (cross-flow)	* 0.58–0.61	[47]
Heating coil in the module	LHIF	Red seawater	1 kW	9.8	1.18	[48]
Aluminum layer	SHVMD-3	35 g/L NaCl	n.a.	9	1.17	[49]
Aluminum layer and aluminum meshes	VMD	100 g/L NaCl	25 W	7.6	0.87	[50]
Pt-MBT@Ag NSs/NF spacer	DCMD	0.5 M NaCl	0.8	3.6	2.5	[51]
Pt-Ni foam spacer	DCMD	5 g/L NaCl	50 W	13	2.8 ** a	[52]
P-G-Ni$_{foam}$ light absorber	SVGMD	3.25–16.70 wt% NaCl Oil-contaminated water	1 1	1.13–0.96 1.07	* 0.88–1.04 * 0.93	[53]

* calculated as the ratio between the Energy Supply for localized heating (kW·m^{-2}) and the Flux (kg·m^{-2} h^{-1}).
** heater input energy per produced distillate. a external heated feed.

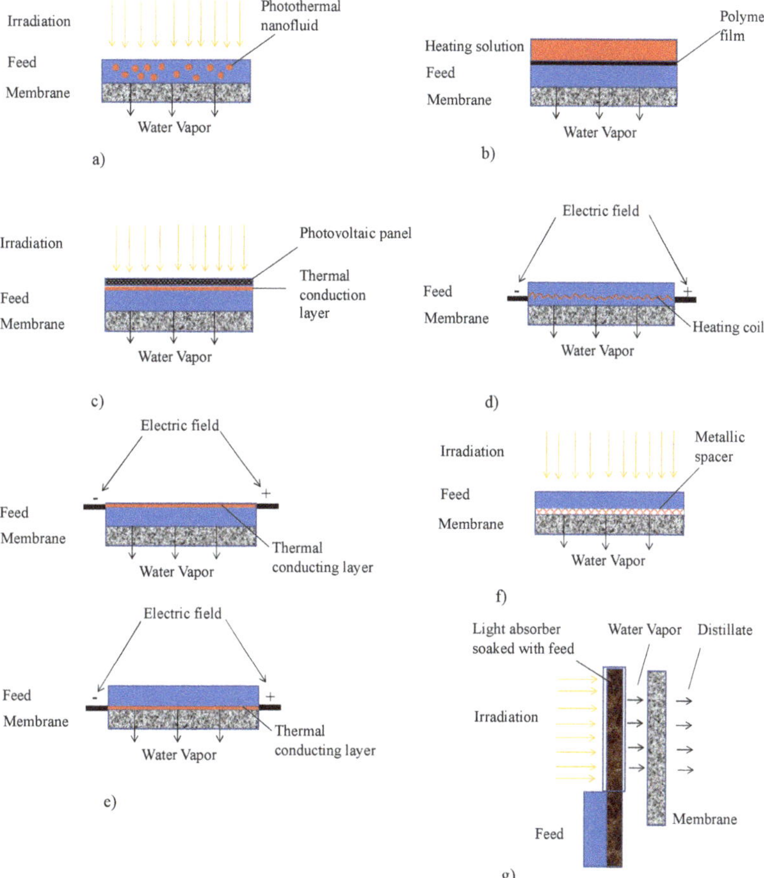

Figure 4. Investigated approaches for heating the feed inside the module through the use of (**a**) photothermal nanofluids, (**b**) a heating solution, (**c**) solar cells, (**d**) heating coils, (**e**) thermal conducting layers, (**f**) metallic spacers and (**g**) light absorbers.

Han et al. [49] used aluminum shim as thermal conducting layer in VMD desalination. Three different locations of the shim inside the cell were investigated: into the feed channel (SHVMD-1), close to the membrane surface (SHVMD-2), and both into the feed channel and close to the membrane (SHVMD-3). A hydrophobic PTFE membrane with an effective area of 40 cm^2 was used. The SHVMD-2 and SHVMD-3 designs were proven to be the most efficient ones, leading to flux values of 7 kg·m^{-2} h^{-1} and 9 kg·m^{-2} h^{-1}, respectively, with corresponding SEC values of 1.1 kW·kg^{-1} and 1.17 kW·kg^{-1}. A reduction of 20% in SEC at the expense of less than 10% flux reduction can be obtained through intermittent heating.

Aluminum layer and aluminum meshes were located inside a module equipped with a PP membrane (40 cm^2), to supply localized heat, by Wang et al. [50]. VMD tests were carried out on a salty feed with different direct heating configurations: aluminum layer in the feed channel, not in contact with the membrane; aluminum meshes in direct contact with the membrane (at the feed or at the distillate side); aluminum layer in the feed channel and aluminum meshes at the distillate side. The highest flux was obtained by combining the aluminum layer and meshes (7.6 kg·m^{-2} h^{-1}) with an SEC of 0.87 kW·kg^{-1}, while the lowest SEC was achieved by the mesh-only configuration (0.26 kW·kg^{-1}) which led to a flux of 3.5 kg·m^{-2} h^{-1}.

In order to overcome the issues of the membrane pore reduction and of the NPs loss under feed recirculation, which are usually encountered when modifying the membrane to heat it under solar irradiation, Ang et al. [51], created a plasmonic spacer to be used in the module. It consisted of Pt NPs grown on a porous nickel foam (NF), an Ag NP coating, and a nano-ligand (4,8-bis(methylthio)benzo [1,2-d:4,5-d0]bis [1,3]dithiole-2,6-dithione, denoted as MBT) embedded at the Pt–Ag interface (Pt-MBT@Ag NSs/NF). The spacer was located in direct contact with the membrane at the feed side and led to a 70% increase of the distillate flux with respect to the pristine NF (3.6 vs. 2.1 kg·m^{-2} h^{-1}), with a photothermal efficiency of 98% under 0.8 kW·kg^{-1} irradiation.

Metal spacers (Ni and Cu foams) also coated with platinum nanosheets (Pt NSs) photocatalyst were employed at the feed side of a membrane module (37.1 cm^2) made of clear acrylic, so as to allow the irradiation of the feed channel, by Tan et al. [52]. Tests were carried out on a salty solution in DCMD mode with a PVDF membrane. During experiments, the feed was recirculated at 65 °C using a hot plate stirrer. All metal spacers led to similar fluxes of the traditional PP spacer. However, the heater input energy per unit volume distillate of the PP spacer was the highest (4 kW·kg^{-1}), while with the metallic foams lower values were achieved (up to 21% lower, under irradiation, because of the absorption of the heat from the light source), and further reduced (by 28%) when the Pt-coated Ni foam spacer was used due to the photothermal conversion.

Gong et al. [53] showed a particular solar vapor gap membrane distillation (SVGMD) process where a free-standing graphene-nickel foam with polymer coating (P–G–Ni$_{foam}$) was prepared and used to transport (via graphene nanochannels) the feed from a sink where it was immersed. The feed was then evaporated under localized solar irradiation. The produced vapor migrated through a gap and reached a PVDF membrane, which blocked all microorganisms at one side and led to the final distillate at the other side. With this system, the direct contact of the feed and the membrane was prevented, so fouling issues were avoided. Tests were carried out on salty solutions at lab scale (membrane area 4 cm^2) and on oil/water mixture in a scaled-up system (membrane area 21 cm^2). In the second application, while the feed water was lifted to the upper part, the oil was rejected underwater, thanks to the superhydrophilic and underwater superoleophobic nature of the light absorber. A high solar efficiency (73.4% at 1 kW·m^{-2}) was obtained during tests on salty water. When feeding oil-contaminated water, the TOC concentration in the distillate was <2 mg/L, confirming that oil was effectively rejected. Obtained fluxes are listed in Table 3.

Some Remarks

The heating of the fluid inside the module was obtained by following different strategies, with the final aim of reducing the energy needed to heat the feed (no external heating systems were used, except in one case study) and of increasing the temperature close to the membrane surface, without acting on the membrane itself. PVDF and PTFE membranes were mostly used and MD tests on salty solutions always rendered high-purity permeates. The research addressed the employment of photothermal nanofluids, heating solutions, solar cells, coils, and thermal conductive layers, as well as plasmonic spacers. A particular system where a light absorber material was used in combination with a PVDF membrane was also investigated. Typical membrane areas were in the range of 4–200 cm^2, with modules up to 8 m^2 developed when using heating solutions. As for the irradiated modified membranes, in cases where photothermal elements are used inside the module, a window for the irradiation must be designed at the feed side. When compared to conventional MD modules, better performances were achieved. Flux and SEC ranged from 0.74 kg·m^{-2} h^{-1} to 13 kg·m^{-2} h^{-1} and from 0.58 kW·kg^{-1} to 2.8 kW·kg^{-1}, respectively.

5. Concluding Remarks and Future Perspectives

Localized heating was investigated as a means of improving the thermal efficiency of membrane distillation. In this respect, actions were made on the membrane properties and on the module design. Both solar and electric energy were considered for the energy supply and, as a general result, improvements in trans-membrane flux and a reduction of energy consumptions were achieved with respect to traditional MD systems. For example, a comparison of the specific energy consumption of a localized heated MD, where a silver membrane connected with an electrical source was used, and those obtained in conventional MD configurations led to a significant difference (from 10,000–30,000 kJ·kg^{-1} of evaporate for localized heating versus 60,000–120,000 kJ·kg^{-1} for traditional DCMD and 57,600–122,400 kJ·kg^{-1} for conventional AGMD) [54]. Interestingly, due to heating close to the membrane surface in the investigated systems, the TPF could reach values higher than 100%, while in conventional MD it is often significantly lower. However, the higher temperature at the membrane surface could lead to the deterioration of thermolabile molecules (e.g., proteins), which, when present in the feed need to be treated. Therefore, it must be managed with care. To the best of our knowledge, no studies of localized heating for treating feeds containing thermolabile compounds were carried out until now. Nevertheless, this aspect deserves particular attention if the localized heating approach is going to be applied in the food and beverage or pharmaceutical fields. Plasmonic spacers and modified membranes, with enhanced hydrophobicity and antifouling properties, were successfully prepared and tested mainly at lab scale. The use of solar energy is certainly a sustainable choice; however, MD performance could be affected by an unstable freshwater production during periods with low solar irradiation. Membranes with good photothermal and Joule heating proved to be an interesting option to ensure a constant productivity, while reducing the electrical energy consumptions linked to the Joule heating. On the other hand, Joule heating could cause water splitting and membrane degradation in high-salinity environments, thus, isolation of the electrothermal material from saline water must be considered. It has to be noted that when modifications are carried out on membranes/spacers, an aspect of concern is the possible release in time of the heating materials used, with a consequent reduction of the system efficiency together with a pollution of the feed stream. In addition, the life time of these materials must be investigated, as well as the most appropriate strategies to reduce fouling issues without affecting the membranes/spacers properties. Commercial membranes without modification were also tested when the localized heating was made inside the module. Both localized heating strategies (acting on the membrane or acting on the module) increase the complexity of the MD plant, due to the need of specifically designed modules (especially for solar irradiation tests) and of systems for an efficient energy supply, like external electric circuits or the use of a lens to collect and concentrate solar energy. Moreover, for modules to be used under irradiation, flat membranes must be employed, with consequent reduction of the module compactness. Therefore, although encouraging results were obtained in both localized heating strategies, it is too early to make a clear choice among the tested MD units. Research is still at the first stage and further optimizations are needed in terms of type and amount of material to be used in membrane/spacer modification, power intensity, inclination of the unit to enhance solar irradiation, feed flow rate, etc. For instance, it was demonstrated that better efficiencies can be obtained without re-circulating the feed, so as to reduce the energy consumption for heating and the heat losses along the circuit. However, strategies to minimize fouling must be conceived (e.g., intermittent flushing). Moreover, lower energy consumptions were obtained with heat recovery inside the module. In addition to the above observations, for a large-scale implementation of the modified membranes and plasmonic spacers, it is also important to ensure their stability in time (long-term tests are needed) and to evaluate their ease in upscaling. Furthermore, the scaling up of the specifically designed module must also be carried out, as well as the development of systems to control the input power by artificial intelligence technologies. Therefore, there are various aspects which need further investigation before a localized heating MD can be adopted. Nevertheless, it is expected that the localized heating approach will significantly impact the application of membrane distillation in different fields.

Author Contributions: Conceptualization, A.C.; supervision, A.C.; writing—original draft preparation, A.C. and M.C.C.; writing—review and editing, A.C. All authors have read and agreed to the published version of the manuscript.

Funding: This research received no external funding.

References

1. Wang, P.; Chung, T.-S. Recent advances in membrane distillation processes: Membrane development, configuration design and application exploring. *J. Membr. Sci.* **2015**, *474*, 39–56. [CrossRef]
2. Thomas, N.; Mavukkandy, M.O.; Loutatidou, S.; Arafat, H.A. Membrane distillation research & implementation: Lessons from the past five decades. *Sep. Purif. Technol.* **2017**, *189*, 108–127. [CrossRef]
3. Deshmukh, A.; Boo, C.; Karanikola, V.; Lin, S.; Straub, A.P.; Tong, T.; Warsinger, D.M.; Elimelech, M. Membrane distillation at the water-energy nexus: Limits, opportunities, and challenges. *Energy Environ. Sci.* **2018**, *11*, 1177–1196. [CrossRef]
4. Lee, H.Y.; He, F.; Song, L.M.; Gilron, J.; Sirkar, K.K. Desalination with a cascade of cross-flow hollow fiber membrane distillation devices integrated with a heat exchanger. *AICHE J.* **2011**, *57*, 1780–1795. [CrossRef]
5. Lin, S.; Yip, N.Y.; Elimelech, M. Direct contact membrane distillation with heat recovery: Thermodynamic insights from module scale modeling. *J. Membr. Sci.* **2014**, *453*, 498–515. [CrossRef]
6. Guan, G.; Yang, X.; Wang, R.; Fane, A.G. Evaluation of heat utilization in membrane distillation desalination system integrated with heat recovery. *Desalination* **2015**, *366*, 80–93. [CrossRef]
7. Winter, D.; Koschikowski, J.; Wieghaus, M. Desalination using membrane distillation: Experimental studies on full scale spiral wound modules. *J. Membr. Sci.* **2011**, *375*, 104–112. [CrossRef]
8. Zhao, K.; Heinzl, W.; Wenzel, M.; Buttner, S.; Bollen, F.; Lange, G.; Heinzl, S.; Sarda, N. Experimental study of the memsys vacuum-multi-effect-membrane-distillation (V-MEMD) module. *Desalination* **2013**, *323*, 150–160. [CrossRef]
9. Jansen, A.E.; Assink, J.W.; Hanemaaijer, J.H.; van Medevoort, J.; van Sonsbeek, E. Development and pilot testing of full-scale membrane distillation modules for deployment of waste heat. *Desalination* **2013**, *323*, 55–65. [CrossRef]
10. Mohamed, E.S.; Boutikos, P.; Mathioulakis, E.; Belessiotis, V. Experimental evaluation of the performance and energy efficiency of a Vacuum Multi-Effect Membrane Distillation system. *Desalination* **2017**, *408*, 70–80. [CrossRef]
11. Tamburini, A.; Pitò, P.; Cipollina, A.; Micale, G.; Ciofalo, M. A thermochronic liquid crystals image analysis technique to investigate temperature polarization in spacer-filled channels for membrane distillation. *J. Membr. Sci.* **2013**, *447*, 260–273. [CrossRef]
12. Taamneh, Y.; Bataineh, K. Improving the performance of direct contact membrane distillation utilizing spacer-filled channel. *Desalination* **2017**, *408*, 25–35. [CrossRef]
13. Soukane, S.; Nacuer, M.W.; Francis, L.; Alsaadi, A.; Ghaffour, N. Effect of feed flow pattern on the distribution of permeate fluxes in desalination by direct contact membrane distillation. *Desalination* **2017**, *418*, 43–59. [CrossRef]
14. Criscuoli, A. Experimental investigation of the thermal performance of new flat membrane module designs for membrane distillation. *Int. Comm. Heat Mass Transf.* **2019**, *103*, 83–89. [CrossRef]
15. Summers, E.K.; Arafat, H.A.; Lienhard, J.H.V. Energy efficiency in comparison of single-stage membrane distillation (MD) desalination cycles in different configurations. *Desalination* **2012**, *290*, 54–66. [CrossRef]
16. Ye, H.; Li, X.; Deng, L.; Li, P.; Zhang, T.; Wang, X.; Hsiao, B.S. Silver nanoparticle-enabled photothermal nanofibrous membrane for light-driven membrane distillation. *Ind. Eng. Chem. Res.* **2019**, *58*, 3269–3281. [CrossRef]
17. Song, L.; Huang, Q.; Huang, Y.; Bi, R.; Xiao, C. An electro-thermal braid-rein- forced PVDF hollow fiber membrane for vacuum membrane distillation. *J. Membr. Sci.* **2019**, *591*, 117359. [CrossRef]
18. Zuo, K.; Wang, W.; Deshmukh, A.; Jia, S.; Guo, H.; Xin, R.; Elimelech, M.; Ajayan, P.M.; Lou, J.; Li, Q. Multifunctional nanocoated membranes for high-rate electrothermal desalination of hypersaline waters. *Nat. Nanotechnol.* **2020**, *15*, 1025–1032. [CrossRef]
19. Dudchenko, A.V.; Chen, C.; Cardenas, A.; Rolf, J.; Jassby, D. Frequency-dependent stability of CNT Joule heaters in ionizable media and desalination processes. *Nat. Nanotechnol.* **2017**, *12*, 557–563. [CrossRef]
20. Huang, J.; Tang, T.; He, Y. Coupling photothermal and Joule-heating conversion for self-heating membrane distillation enhancement. *Appl. Therm. Eng.* **2021**, *199*, 117557. [CrossRef]
21. Huang, J.; He, Y.; Shen, Z. Joule heating membrane distillation enhancement with multi-level thermal concentration and heat recovery. *Energy Convers. Manag.* **2021**, *238*, 114111. [CrossRef]
22. Ahmed, F.E.; Lalia, B.S.; Hashaikeh, R.; Hilal, N. Enhanced performance of direct contact membrane distillation via selected electrothermal heating of membrane surface. *J. Membr. Sci.* **2020**, *610*, 118224. [CrossRef]
23. Li, K.; Zhang, Y.; Wang, Z.; Liu, L.; Liu, H.; Wang, J. Electrothermally Driven Membrane Distillation for Low-Energy Consumption and Wetting Mitigation. *Environ. Sci. Technol.* **2019**, *53*, 13506–13513. [CrossRef]

24. Subrahmanya, T.M.; Lin, P.T.; Chiao, Y.-H.; Widakdo, J.; Chuang, C.-H.; Rahmadhanty, S.F.; Yoshikawa, S.; Hung, W.-S. High Performance Self-Heated Membrane Distillation System for Energy Efficient Desalination Process. *J. Mater. Chem. A* **2021**, *9*, 7868–7880. [CrossRef]

25. Shukla, S.; Méricq, J.P.; Belleville, M.P.; Hengl, N.; Benes, N.E.; Vankelecom, I.; Sanchez Marcano, J. Process intensification by coupling the Joule effect with pervaporation and sweeping gas membrane distillation. *J. Membr. Sci.* **2018**, *545*, 150–157. [CrossRef]

26. Anvari, A.; Kekre, K.M.; Azimi, Y.A.; Yao, Y.; Ronen, A. Membrane distillation of high salinity water by induction heated thermally conducting membranes. *J. Membr. Sci.* **2019**, *589*, 117253. [CrossRef]

27. Politano, A.; Argurio, P.; Di Profio, G.; Sanna, V.; Cupolillo, A.; Chakraborty, S.; Arafat, H.A.; Curcio, E. Photothermal Membrane Distillation for Seawater Desalination. *Adv. Mater.* **2017**, *29*, 1603504. [CrossRef]

28. Politano, A.; Di Profio, G.; Fontananova, E.; Sanna, V.; Cupolillo, A.; Curcio, E. Overcoming temperature polarization in membrane distillation by thermoplasmonic effects activated by Ag nanofillers in polymeric membranes. *Desalination* **2019**, *451*, 192–199. [CrossRef]

29. Pagliero, M.; Alloisio, M.; Costa, C.; Firpo, R.; Mideksa, E.A.; Comite, A. Carbon Black/Polyvinylidene Fluoride Nanocomposite Membranes for Direct Solar Distillation. *Energies* **2022**, *15*, 740. [CrossRef]

30. Dongare, P.D.; Alabastri, A.; Pedersen, S.; Zodrow, K.R.; Hogan, N.J.; Neumann, O.; Wu, J.; Wang, T.; Deshmukh, A.; Elimelech, M.; et al. Nanophotonics-enabled solar membrane distillation for off-grid water purification. *Proc. Natl. Acad. Sci. USA* **2017**, *114*, 6936–6941. [CrossRef]

31. Said, I.A.; Wang, S.; Li, Q. Field Demonstration of a Nanophotonics-Enabled Solar Membrane Distillation Reactor for Desalination. *Ind. Eng. Chem. Res.* **2019**, *58*, 18829–18835. [CrossRef]

32. Wu, J.; Zodrow, K.R.; Szemraj, P.B.; Li, Q. Photothermal nanocomposite membranes for direct solar membrane distillation. *J. Mater. Chem. A* **2017**, *5*, 23712–23719. [CrossRef]

33. Tanvir, R.U.; Sujon, S.A.; Yi, P. Passive Permeate-Side -Heated Solar Thermal Membrane Distillation: Extracting Potable Water from Seawater, Surface Water, and Municipal Wastewater at High Single-Stage Solar Efficiencies. *ACS EST Engg.* **2021**, *1*, 770–779. [CrossRef]

34. Chen, Y.R.; Xin, R.; Huang, X.; Zuo, K.; Tung, K.L.; Li, Q. Wetting-resistant photothermal nanocomposite membranes for direct solar membrane distillation. *J. Membr. Sci.* **2021**, *620*, 118913. [CrossRef]

35. Gong, B.; Yang, H.; Wu, S.; Yan, J.; Cen, K.; Bo, Z.; Ostrikov, K.K. Superstructure-Enabled anti-fouling membrane for efficient photothermal distillation. *ACS Sustain. Chem. Eng.* **2019**, *7*, 20151–20158. [CrossRef]

36. Wu, X.; Jiang, Q.; Ghim, D.; Singamaneni, S.; Jun, Y.-S. Localized heating with a photothermal polydopamine coating facilitates a novel membrane distillation process. *J. Mater. Chem. A* **2018**, *6*, 18799–18807. [CrossRef]

37. Ghim, D.; Wu, X.; Suazo, M.; Jun, Y.S. Achieving Maximum Recovery of Latent Heat in Photothermally Driven Multi-Layer Stacked Membrane Distillation. *Nano Energy.* **2021**, *80*, 105444. [CrossRef]

38. Huang, Q.; Gao, S.; Huang, Y.; Zhang, M.; Xiao, C. Study on photothermal PVDF/ATO nanofiber membrane and its membrane distillation performance. *J. Membr. Sci.* **2019**, *582*, 203–210. [CrossRef]

39. Zhang, Y.; Li, K.; Liu, L.; Wang, K.; Xiang, J.; Hou, D.; Wang, J. Titanium nitride nanoparticle embedded membrane for photothermal membrane distillation. *Chemosphere* **2020**, *256*, 127053. [CrossRef]

40. Li, W.; Chen, Y.; Yao, L.; Ren, X.; Li, Y.; Deng, L. Fe_3O_4/PVDFHFP Photothermal Membrane with in-Situ Heating for Sustainable, Stable and Efficient Pilot-Scale Solar-Driven Membrane Distillation. *Desalination* **2020**, *478*, 114288. [CrossRef]

41. Huang, J.; Hu, Y.; Bai, Y.; He, Y.; Zhu, J. Novel solar membrane distillation enabled by a PDMS/CNT/PVDF membrane with localized heating. *Desalination* **2020**, *489*, 114529. [CrossRef]

42. Han, X.; Wang, W.; Zuo, K.; Chen, L.; Yuan, L.; Liang, J.; Li, Q.; Ajayan, P.M.; Zhao, Y.; Lou, J. Bio-derived ultrathin membrane for solar driven water purification. *Nano Energy* **2019**, *60*, 567–575. [CrossRef]

43. Tan, Y.Z.; Wang, H.; Han, L.; Tanis-Kanbur, M.B.; Pranav, M.V.; Chew, J.W. Photothermal-enhanced and fouling-resistant membrane for solar-assisted membrane distillation. *J. Membr. Sci.* **2018**, *565*, 254–265. [CrossRef]

44. Mustakeem, M.; El-Demellawi, J.K.; Obaid, M.; Fangwang, M.; Alshareef, H.N. MXene-Coated Membranes for Autonomous Solar-Driven Desalination. *ACS Appl. Mater.* **2022**, *14*, 5265–5274. [CrossRef] [PubMed]

45. Zhang, Y.; Liu, L.; Li, K.; Hou, D.; Wang, J. Enhancement of Energy Utilization Using Nanofluid in Solar Powered Membrane Distillation. *Chemosphere* **2018**, *212*, 554–562. [CrossRef] [PubMed]

46. Schwantes, R.; Seger, J.; Bauer, L.; Winter, D.; Hogen, T.; Koschikowski, J.; Geißen, S.U. Characterization and Assessment of a Novel Plate and Frame MD Module for Single Pass Wastewater Concentration–FEED Gap Air Gap Membrane Distillation. *Membranes* **2019**, *9*, 118. [CrossRef]

47. Wang, W.; Shi, Y.; Zhang, C.; Hong, S.; Shi, L.; Chang, J.; Li, R.; Jin, Y.; Ong, C.; Zhuo, S.; et al. Simultaneous production of fresh water and electricity via multistage solar photovoltaic membrane distillation. *Nat. Commun.* **2019**, *10*, 3012. [CrossRef]

48. Mustakeem, M.; Qamar, A.; Alpatova, A.; Ghaffour, N. Dead-end membrane distillation with localized interfacial heating for sustainable and energy-efficient desalination. *Water Res.* **2021**, *189*, 116584. [CrossRef]

49. Han, F.; Liu, S.; Wang, K.; Zhang, X. Enhanced Performance of Membrane Distillation Using Surface Heating Process. *Membranes* **2021**, *11*, 866. [CrossRef]
50. Wang, J.; Liu, Y.; Rao, U.; Dudley, M.; Ebrahimi, N.D.; Lou, J.; Han, F.; Hoek, E.M.V.; Tilton, N.; Cath, T.Y.; et al. Conducting thermal energy to the membrane/water interface for the enhanced desalination of hypersaline brines using membrane distillation. *J. Membr. Sci.* **2021**, *626*, 119188. [CrossRef]
51. Ang, E.H.; Tan, Y.Z.; Chew, J.W. A three-dimensional plasmonic spacer enables highly efficient solar-enhanced membrane distillation of seawater. *J. Mater. Chem. A* **2019**, *7*, 10206. [CrossRef]
52. Tan, Y.Z.; Ang, E.H.; Chew, J.W. Metallic spacers to enhance membrane distillation. *J. Membrane Sci.* **2019**, *572*, 171–183. [CrossRef]
53. Gong, B.; Yang, H.; Wu, S.; Xiong, G.; Yan, J.; Cen, K.; Bo, Z.; Ostrikov, K. Graphene Array-Based Anti-fouling Solar Vapour Gap Membrane Distillation with High Energy Efficiency. *Nano-Micro Lett.* **2019**, *11*, 51. [CrossRef] [PubMed]
54. Alsaati, A.; Marconnet, A.M. Energy efficient membrane distillation through localized heating. *Desalination* **2018**, *442*, 99–107. [CrossRef]

Article

Carbon Black/Polyvinylidene Fluoride Nanocomposite Membranes for Direct Solar Distillation

Marcello Pagliero [1,2,*], Marina Alloisio [1], Camilla Costa [1,2], Raffaella Firpo [1,2], Ermias Ararsa Mideksa [1] and Antonio Comite [1,2]

[1] Department of Chemistry and Industrial Chemistry, University of Genoa, Via Dodecaneso 31, 16146 Genova, Italy; marina.alloisio@unige.it (M.A.); camilla.costa@unige.it (C.C.); raffaella.firpo@unige.it (R.F.); ermiasararsa@gmail.com (E.A.M.); antonio.comite@unige.it (A.C.)
[2] INSTM, Udr di Genova, Via Dodecaneso 31, 16146 Genova, Italy
* Correspondence: marcello.pagliero@unige.it

Abstract: Water reclamation is becoming a growing need, in particular in developing countries where harvesting the required energy can be a challenging problem. In this context, exploiting solar energy in a specifically tailored membrane distillation (MD) process can be a viable solution. Traditional MD guarantees a complete retention of non-volatile compounds and does not require high feed water temperatures. In this work, a suitable amount of carbon black (CB) was incorporated into the whole matrix of a polymeric porous membrane in order to absorb light and directly heat the feed. The mixed matrix membranes were prepared forming a uniform CB dispersion in the PVDF dope solution and then using a non-solvent induced phase separation process, which is a well-established technique for membrane manufacturing. CB addition was found to be beneficial on both the membrane structure, as it increased the pore size and porosity, and on the photothermal properties of the matrix. In fact, temperatures as high as 60 °C were reached on the irradiated membrane surface. These improvements led to satisfactory distillate flux (up to 2.3 L/m^{2}h) during the direct solar membrane distillation tests performed with artificial light sources and make this membrane type a promising candidate for practical applications in the field of water purification.

Keywords: carbon black; photothermal; direct solar membrane distillation; PVDF; renewable energy

Citation: Pagliero, M.; Alloisio, M.; Costa, C.; Firpo, R.; Mideksa, E.A.; Comite, A. Carbon Black/ Polyvinylidene Fluoride Nanocomposite Membranes for Direct Solar Distillation. *Energies* **2022**, *15*, 740. https://doi.org/ 10.3390/en15030740

Academic Editor: Alessandra Criscuoli

Received: 17 December 2021
Accepted: 17 January 2022
Published: 20 January 2022

Publisher's Note: MDPI stays neutral with regard to jurisdictional claims in published maps and institutional affiliations.

1. Introduction

Membrane distillation (MD) is a thermally driven separation process based on a hydrophobic porous membrane which contacts a hot concentrated solution and pure water. The temperature difference applied at each side of the membrane induces a vapor pressure difference across the membrane that acts as driving force for the process and establishes a water vapor flux from the hot to the cold section [1,2].

Membranes used for MD application must comply with important specific requirements. First, high porosity is mandatory to maximize the trans-membrane water flux since the mass transfer coefficient is proportional to the mean porosity of the membrane itself [3]. Another essential feature is good surface hydrophobicity. In fact, the pores need to remain dry during MD operation since their flooding can cause a decrease of the distillate flux, as well as a reduction of the separation ability of the membrane [4,5]. This property can be controlled either by decreasing the surface wettability or by developing membranes with small pores and a narrow pore size distribution. Smaller pores are indeed less prone to wetting phenomena [6].

MD membranes exist in two basic configurations: hollow fiber and flat sheet. They can be arranged in different modules [7,8] and are generally prepared using hydrophobic polymers, such as polypropylene [9,10], polyethylene [11], polytetrafluoroethylene (PTFE) [12,13] and polyvinylidene fluoride (PVDF) [14–17], or using modified ceramic materials [18–20]. Among all of the aforementioned materials, PVDF is particularly attractive because it can be easily dissolved in most common organic solvents so membranes can be prepared via traditional processes, such as thermally induced phase separation (TIPS) and non-solvent induced phase separation (NIPS) [21].

Four main configurations of MD have been developed, namely direct contact MD (DCMD), sweeping gas MD (SGMD), air gap MD (AGMD) and vacuum MD (VMD). The main differences among these configurations reside in the distillate section of the equipment [22].

In this work, VMD configuration has been used to assess the performance of the prepared membranes. In this mode, a partial vacuum pressure is applied on the distillate side of the membrane to remove the water vapor molecules from the membrane module. Since the partial water vapor pressure is constantly maintained as low as possible, VMD guarantees the highest fluxes among all of the MD configurations and reduces the heat losses correlated to heat conduction through the membrane. However, in some conditions, the net pressure difference across the membrane can favor pore flooding, possibly causing a reduction of the process productivity and separation ability [2,23].

Nowadays, MD is a mature technology and it appears to be one of the best technologies for developing off-grid desalination/wastewater treatment plants due to its modular configuration, low power consumption and ability to harness low-grade energy resources [24].

An important problem affecting MD scale up is the heat loss associated to water evaporation in the feed section along the surface of the membrane module. The phase transition that takes place at the feed/pore interface reduces the temperature of water flowing along the module in the feed channel, gradually decreasing the effective driving force of the process [25]. This effect in practice limits the maximum length of the membrane module. An effective solution to contain the feed temperature reduction along the membrane module recently developed is the so-called membrane localized heating that can be obtained using several different techniques. Tan et al. [26], for example, harnessed the high thermal conductivity of nickel and its capacity to be inductively heated by placing a nickel foam in the feed channel of the membrane cell in place of a traditional plastic spacer. The advantage was the possibility of increasing the foam temperature by electromagnetic induction, and therefore heating the feed directly inside the membrane cell. Experimental tests proved that this arrangement was able to improve the energy efficiency and the productivity of the DCMD process. A similar solution was followed by Anvari et al. [27] who spray-coated a commercial PTFE membrane with a layer containing multiwall carbon nanotubes coated with iron oxide. This added layer had magnetic properties that allowed for providing inductive heating directly on the surface of the membrane. It was observed that increasing the nanoparticles loading and the feed residence time improved the distillate flux since heat transfer from the membrane to the feed solution was enhanced. Moreover, these membranes showed high efficiency for feed solutions containing highly saline solutions, proving that this process can be applied to traditional brine treatment.

A particular approach recently explored aims to achieve localized heating by harnessing solar energy. One of the main advantages of the direct solar membrane distillation (DSMD) layout is that the feed temperature is raised in the boundary layer in contact with the membrane surface. Therefore, temperature polarization effects are drastically reduced, and the effective driving force of the process is increased [28]. The DSMD process has been proposed by many researchers as an effective solution to developing larger MD plants [29–31]. The basic solar MD setup consists of a photovoltaic panel used to provide the electrical current required to run the feed water circulation system, and a solar collector that is used to heat the feed. However, this simple configuration is characterized by some flaws, namely the heat losses along the pipes connecting the feed tank to the membrane cell and the need for large solar collectors to reach an adequate feed temperature [32]. Said et al. [28] recently tested a DSMD small pilot plant, assessing its performance in real life operation conditions. The plant was completely powered by solar energy, with a photovoltaic panel providing the electricity for feed circulation and a 0.12 m^2 photoactive membrane cell.

During the testing time the plant mean distillate flux reached 0.55 L/m^{2}h with a rejection factor of 99.8%.

The main research efforts on the development of photoactive membranes for DSMD have been focused at present on the surface modification of commercial hydrophobic membranes; many researchers have coated membrane surfaces with an additional layer containing different fillers, such as carbon nanotubes [33] or carbon black (CB) [28,34–36] and silica-gold nanoparticles [35]. In particular, Wu et al. [35] created a layer composed of polyvinyl alcohol containing carbon black or silica-gold nanospheres. CB-coated membranes showed an increase of 33% in distillate flux when irradiated with solar-like light, while membranes treated with silica-gold nanoparticles produced a more limited improvement of 17%.

In this work, a novel approach was adopted, i.e., hydrophobic photoactive membranes were autonomously prepared in our laboratory, including carbon black directly inside the starting bulk of the polymeric matrix rather than adding a supplementary layer on a preformed membrane. These original nanocomposite membranes were then produced easily with a common NIPS technique. The aim of this study was to explore the effect of the filler on the membrane photothermal properties, as well as on the membrane's internal structure. The influence of CB loading on the MD performance was investigated. To the best of our knowledge, intrinsically photoactive membranes of this particular kind, intended for direct solar membrane distillation, have not been prepared and studied before.

2. Material and Methods

2.1. Dope Solution and Photoactive Membrane Preparation

The membrane type studied in this work was prepared in our laboratory through a method described in a previous article [37]. CB (Vulcan XC72R, Cabot Corp—Boston, MA, USA, primary particle size: 30–60 nm [38]) was first dispersed in a green organic solvent (i.e., triethyl phosphate, TEP, Merck—Darmstadt, Germany). To this end, a precise amount of CB powder was placed inside a 50 mL bottle together with 15 g of TEP and was then sonicated with an ultrasonic bath for 30 min at 25 °C to improve the dispersion.

The membranes were prepared via the NIPS technique, by dissolving a commercial PVDF (Solef® 6010, Solvay Specialty Polymers—Bollate, Italy, Mw 300 kDa) dissolved in the CB/TEP dispersions to create a CB doped solution and provide the desired photothermal properties to the final membrane. A 300 µm thick film of dope solution was cast on a commercial PET non-woven support (Viledon® FO-2401, Freudenberg—Weinheim, Germany) and then immersed in a weak coagulation bath containing ethanol 96 v/v% (VWR International—Radnor, PA, USA). After 2 h, the membranes were removed from the non-solvent, washed with deionized water to remove ethanol from the pores and finally dried overnight at room temperature. Table 1 summarizes the preparation conditions of all of the membranes assessed. In the sample name, the first part refers to the polymer concentration in the dope solution, while the second part refers to the CB concentration.

Table 1. Preparation conditions of the tested membranes.

Sample	145_0	145_2	145_5	145_75
CB concentration [1] [wt%]	0	2.0	5.0	7.5
PVDF concentration [2] [wt%]		14.5		
Solvent		TEP		
Non-solvent		EtOH 96 v/v%		
Casting temperature [°C]		25		
Casting thickness [µm]		300		

[1] with respect to PVDF mass; [2] with respect to the solution mass.

2.2. Light Absorption Measurements

Absolute hemispherical reflectance spectra of the membranes were acquired using an UV/VIS/NIR spectrophotometer (Lambda9, Perkin Elmer—Waltham, MA, USA) equipped with a 150 mm integrating sphere. The inspected spectral range was between 200 and 2700 nm. A TiO$_2$ coated layer was used as a blank sample.

For each sample, four spectra were registered, rotating the specimen by 90° each time. The four spectra allowed for estimating the surface homogeneity of the sample and compensating for any possible influence related to defects in the sample and its surface roughness. The final results were reported as absorbance values, which were then calculated as the mean of the four measurements.

The photothermal properties of the dry membranes were assessed by measuring the surface temperature change over time after a light source was turned on with the setup schematized in Figure 1.

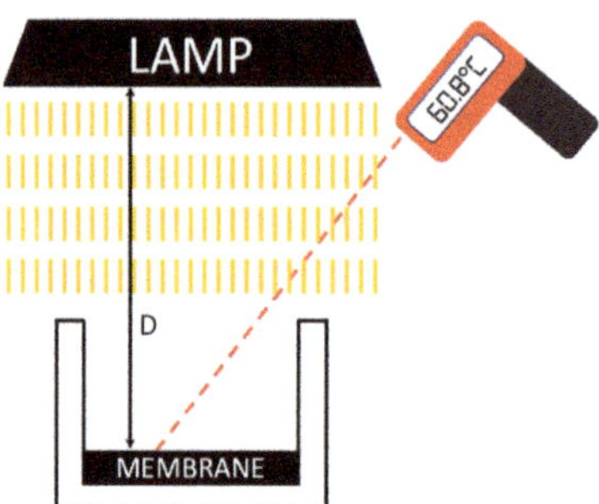

Figure 1. Sketch of the setup used for surface temperature measurements.

The samples (17.5 cm^2 surface area) were put in a special container made of expanded PVC—used to avoid heat losses related to conduction through the membrane—and placed under the light source at a distance (D) of 9 cm. The surface temperature was determined using an infrared thermometer (RS-8662, RS PRO—Kuala Lumpur, Malaysia).

2.3. Membrane Performance Evaluation

The prepared membranes were tested using a DSMD setup expressly built in the laboratory for this application. VMD configuration was selected because of the high distillate flux that it guarantees in comparison to other MD modes. Figure 2 reports a scheme of this apparatus.

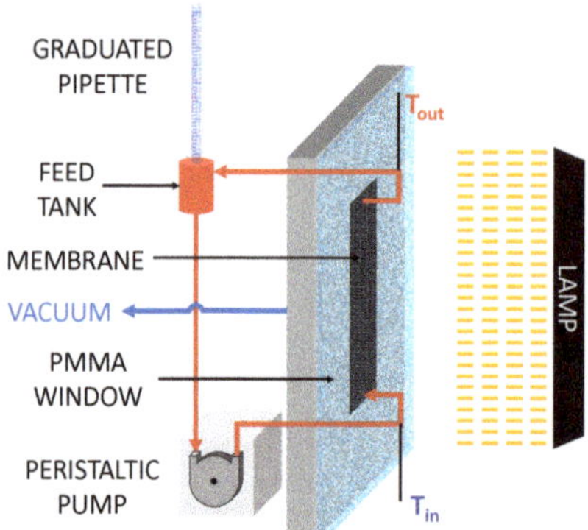

Figure 2. Sketch of the DSMD setup used to test the photothermal membranes.

The membrane cell was equipped with a transparent polymethylmethacrylate (PMMA) film that allowed the passage of simulated solar light. The feed channel was 3 mm thick and the peristaltic pump used for recirculating the liquid had a flow rate of 40 mL/min, therefore, the velocity inside the membrane cell was 53 cm/min and the residence time was almost 15 s. The concentrate was recirculated to the feed tank continuously. Two K type thermocouples, connected to a digital data logger (HD2128.1, Deltaohm—Selvazzano Dentro, Italy), were mounted at the feed inlet and at the concentrate outlet to measure the liquid temperature gradient along the membrane module.

The driving force was applied by connecting the support side of the cell (grey in Figure 2) to a vacuum pump that maintained the permeate side at 20 mbar (absolute pressure). Before the start of each test, a precise amount of deionized water (40 mL) was put in a closed tank equipped with a graduated pipette that was used to measure the distillate flux through the lowering of the liquid level after a set time interval (60 min).

The only heat source for these tests was a light source mounted 7 cm away from the membrane surface. Each membrane was tested using two different light sources: a traditional 75 W incandescent light bulb (OSRAM—Munich, Germany), and a 100 W solar spectrum LED light (Shenzhen Milyn Technology—Guangdong, China). A reference test was also performed in order to evaluate the membrane performance without any external light source. In this case, the membrane cell was covered with an aluminum foil during operation.

Figure 3 reports the emission spectra of the two lamps used, compared with sunlight.

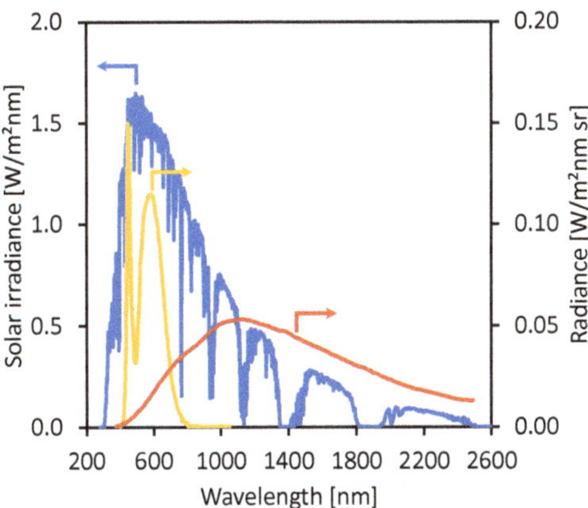

Figure 3. ASTMG 173 AM1.5 solar emission spectrum (blue), and common incandescent light bulb (red) and natural white 100 W LED lamp (yellow) spectra [39]. Note the different ordinate scales.

The irradiance of the two light sources at different distances ranging between 4 cm and approximately 40 cm was determined using a solar power meter (ISM 410, RS PRO—Kuala Lumpur, Malaysia). The measurements were performed by placing the lamp vertically above the sensor.

Table 2 summarizes the operating conditions used to test the DSMD performance of the prepared membranes.

Table 2. DSMD tests conditions.

Tested Feeds	Deionized Water
Light distance	7 cm
Feed temperature	24 °C
Feed flow rate	40 mL/min
Membrane area	30 cm^2
Vacuum degree	20 mbar

3. Results and Discussion

3.1. Light Source Characterization

A major system characteristic to be investigated when using light sources to simulate the solar energy is the irradiated power per unit area, which is called irradiance. The irradiance then must be compared to the average yearly standard solar conditions at the sea level (about 1000 W/m^2) [40]. Figure 4 reports the values collected at different distances for the two lamps employed in this work.

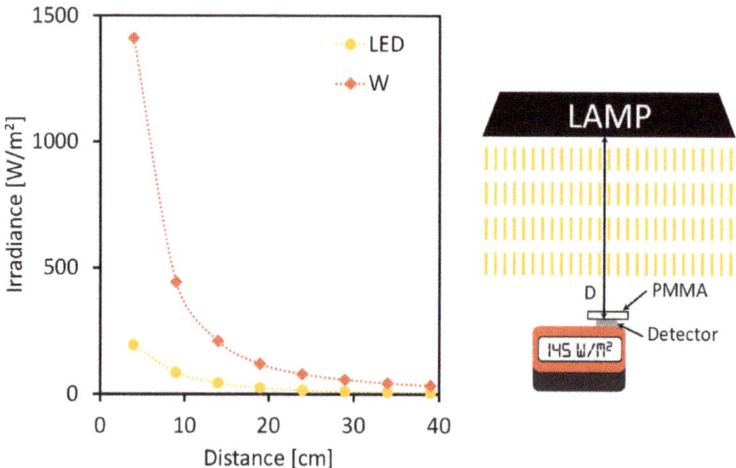

Figure 4. Irradiance of the two used lamps, LED and incandescent, W, through a PMMA window (**left**) and measuring setup scheme (**right**).

The traditional incandescent lamp (red) was able to provide higher power than the solar-like LED lamp (yellow). However, both light sources showed a hyperbolic-like decrease of the irradiance value as the distance between the source and the sensor was increased. This imposed a strict control on the distance between the lamp and the membrane surface in the successive VMD tests.

3.2. Light Absorption of Membranes

Common cells for traditional MD processes are generally made of non-transparent materials, such as stainless steel or plastic, owing to their ability to withstand the feed temperature and salinity. However, such materials are not transparent to UV and visible radiation. Since membranes for DSMD application must be able to absorb light and to transform the energy captured in the available heat, their surface should be allowed to be irradiated with light. Among all of the transparent material, PMMA was selected for this purpose during the VMD cell assembling due to its low absorbance in a broad wavelength interval, as reported in Figure 5.

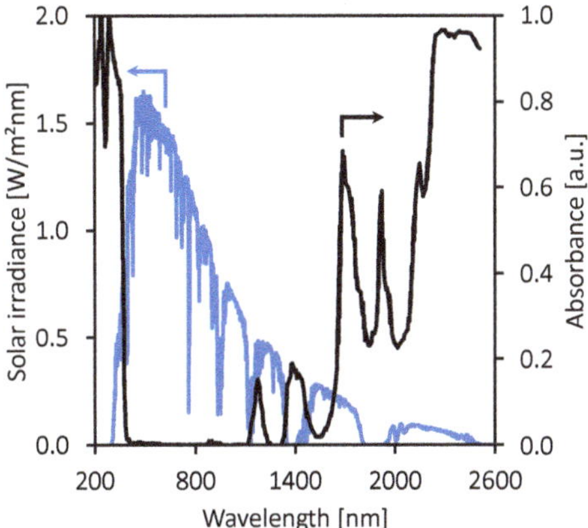

Figure 5. Light absorbance of the VMD cell window made of PMMA (black) and ASTMG 173 AM1.5 solar emission spectrum (blue).

Spectral solar irradiance was shown and PMMA proved to be completely transparent in the majority of the wavelength interval of solar emission spectrum; it presented intense absorption bands only below 400 nm and over 1300 nm. In the interval 400–1300 nm, almost 90% of the whole solar irradiance is emitted. Therefore, PMMA was considered an ideal material to create the window required in this special VMD cell.

When the membrane surface is irradiated with light, the obvious key parameter for DSMD application is the membrane photoactivity that can be estimated by measuring the absorbance in the UV/visible wavelength interval. These measurements were performed using a Perkin Elmer Lambda9 UV/VIS/NIR spectrometer in the spectral range between 200 and 1400 nm (recall the comments on Figure 5). Figure 6 reports the registered spectra for the four membranes prepared at increasing CB loading as well as for the CB as such.

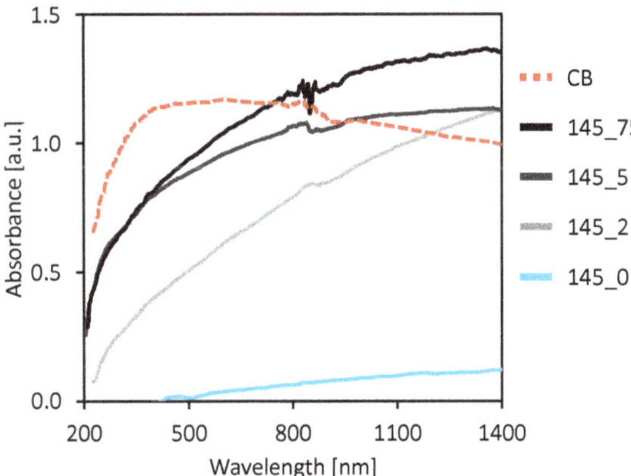

Figure 6. UV/VIS/NIR absorption spectra of the four membranes and of carbon black.

CB (red dotted line in Figure 6) showed a broad absorption band in the whole wavelength interval investigated with a wide maximum in the visible light range.

On the other hand, the pure PVDF membrane (blue line in Figure 6) was characterized by a far less intense absorption in the same wavelength interval. Most of the solar radiation evidently was reflected by the white surface of the membrane.

The introduction of CB inside the dope solution greatly increased the light absorption of the samples in the UV/VIS/NIR spectral region of interest. In particular, the absorbance showed an impressive enhancement, moving from 145_0 sample to 145_2 and then 145_5, i.e., with an increase in CB concentration.

Figure 6 illustrates that the unloaded PVDF membrane had a very low ability to capture solar energy with a slight increase at higher wavelengths. The 145_2 sample partly retains the effect of the PVDF matrix, showing an almost linear growth of the absorbance, but, in contrast, it reaches noticeable values at higher wavelengths. The samples 145_5 and 145_75 exhibited a sharp increment of absorbance at lower wavelengths and their spectra appeared quite similar to that of the CB as such. Moreover, there does not seem to be a substantial difference between the behaviors of the two membranes.

These features were reflected in the photothermal activity shown by the various samples. Figure 7 reports the surface temperature change over time when the membrane surface was exposed to the two different lamps (A: LED lamp, B: incandescent lamp).

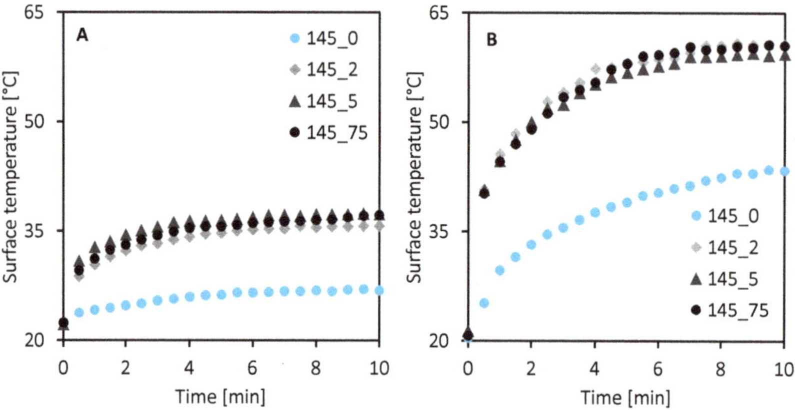

Figure 7. Membrane surface temperature change using (**A**) LED lamp and (**B**) incandescent lamp.

The two lamps were characterized by different emission spectra (Figure 3), as well as a different irradiance (Figure 4). These tests were performed by placing the light source 9 cm above the membrane, meaning that the LED irradiance was around 90 W/m^2, while the incandescent lamp provided about 435 W/m^2 to the membrane surface. This difference was confirmed by the surface temperature measurements. Using the LED light (Figure 7A), the CB loaded membrane's temperature rose almost instantly from about 20 °C to 30 °C and continued to increase for several minutes reaching a stable value at around 35 °C. The incandescent lamp, thanks to its higher irradiance at an equal distance, showed a similar trend but higher temperatures were developed on the membrane surface that reached for all the membranes values of about 60 °C. In both cases A and B, adding CB to the dope solution generated a great difference between unloaded and loaded membrane and an almost 40% temperature increase was obtained for all of the samples. However, the CB concentration had only a marginal effect on the surface temperature of the membrane since, when irradiated under the same conditions, all of the CB-loaded membranes performed similarly. Table 3 reports some literature data of the maximum surface temperature reached by photoactive membranes irradiated with solar-like light sources.

Table 3. Literature data of surface temperature of photoactive membranes.

Membrane Type	Photoactive Material	Irradiation [W/m^2]	Surface Temperature [°C]	Ref.
Coated PVDF	Carbon nanotubes	1000	70	[33]
Coated PVDF	Carbon black	1000	48	[34]
Coated PVDF	Polydopamine	750	35	[41]
Mixed matrix PVDF	Carbon black	435	60	This work

While it is almost impossible to find results obtained under identical conditions, the data reported in Table 3 highlight the excellent performance of the membranes developed in this work. In fact, despite being tested with a less powerful light source, the samples prepared directly including CB inside the PVDF matrix were able to reach temperatures higher than—or at least similar to—the ones obtained with coated membranes.

3.3. Distillation Performance

The membrane distillation performance was evaluated using the setup shown in Figure 2. The feed temperatures were registered over a one-hour test for each light source and the registered differences during this time span, ΔT_m, are reported in Figure 8, and are expressed as follows:

$$\Delta T_m = \frac{\left(T_{out_{60}} - T_{out_0}\right) + \left(T_{in_{60}} - T_{in_0}\right)}{2} \tag{1}$$

where the subscripts in_{60} and out_{60} indicate the liquid temperature at the cell inlet and outlet after 60 min of testing, respectively, while the in_0 and out_0 subscripts represent the same values recorded at the start of the tests.

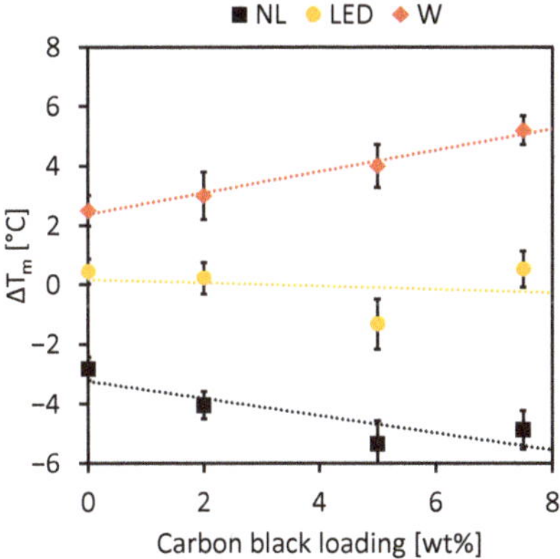

Figure 8. Mean feed temperature difference between MD test end and start with no light (NL, black), LED lamp (yellow) and incandescent light bulb (W, red).

For all of the samples, the test performed without light (black in Figure 8) showed a decrease of the feed temperature over time. In the absence of any external energy supply the latent heat absorbed by water evaporation at the membrane surface caused an appreciable feed cooling.

The ΔT_m values obtained when the membranes were irradiated with a light source exhibited a completely different trend. The LED lamp (yellow in Figure 8) was able to counteract the heat loss related to water evaporation but, since its irradiance was quite weak (see Figure 4), the feed temperature remained almost constant during the test. On the other hand, the incandescent lamp (red in Figure 8) provided more energy to the system and the feed temperature was increased during the time span of DSMD operation for all of the tested samples. These findings are in agreement with the data reported in Figure 7A,B.

The beneficial effect of the light irradiation obviously affected the distillate fluxes as expected. Figure 9 summarizes the results concerning the MD tests for all of the samples and all of the lighting options.

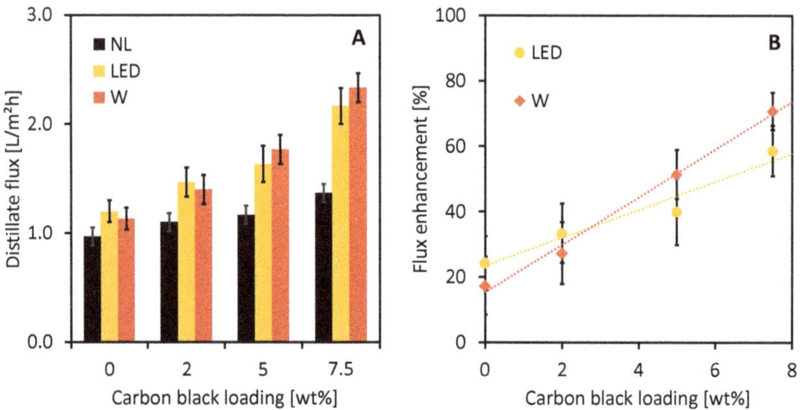

Figure 9. (**A**) mean distillate flux and (**B**) flux enhancement registered during the MD tests with no light (black), LED lamp (yellow) and incandescent light bulb (red).

The distillate fluxes (Figure 9A) obtained during the tests without light sources confirmed the results obtained during a previous work carried out in our laboratory [37]. That study also revealed that the addition of CB to the dope solution enhances both the pore size and porosity of the membrane and therefore can induce an increase in the membrane permeability. It was found that introducing a proper amount of CB led to an increment of both pore size—which passed from about 100 (sample 145_0) to about 850 nm (sample 145_75)—and overall porosity, which rose from 76% to 87%. Both of these parameters strongly affect the membrane mass transfer resistance; it is well known that larger pore size and porosity values provide higher transmembrane vapor fluxes during MD operation. Moreover, in the same study, it was demonstrated that higher CB loadings led to an enhancement of the matrix hydrophobicity, preventing negative phenomena, such as pore flooding. For all of these reasons, even without light irradiation the distillate flux of the 145_75 sample was somewhat higher compared to the one registered for the pure PVDF membrane (145_0).

In this work, using only light as the energy supply for feed heating was proved to be effective for all of the samples prepared. In fact, under irradiation, the distillate flux was always enhanced. This effect became more intense as the CB loading in the dope solution was raised from 2 wt% to 7.5 wt%. The flux enhancement (FE%) reported in Figure 9B was calculated using the following equation:

$$FE\% = \frac{\left(J_{Light} - J_{NL} \right)}{J_{NL}} \cdot 100 \tag{2}$$

where J_{Light} is the distillate flux obtained irradiating the membrane with a light source while J_{NL} is the distillate flux without irradiation.

Increasing the CB concentration improved the light absorbance of the membranes (Figure 6) and boosted the feed temperature rise on the membrane surface. As a consequence, the vapor pressure at the pore entrance got higher and enhanced the driving force of the process. The distillate flux increment became more evident as the CB loading grew. It can be seen from Figure 9B that, for the 145_75 sample, a distillate flux enhancement of up to 60% and 70% was reached when the LED or incandescent lamps were used, respectively. Table 4 reports a VMD performance comparison between the membranes prepared in this work and a sample tested by Ma et al. [42] who coated a PTFE membrane with carbon nanotubes.

Table 4. VMD performance comparison.

Membrane Type	Photoactive Material	Irradiation [W/m^2]	Vacuum [mbar]	Feed Temperature [°C]	Distillate Flux [L/m^{2}h]	Ref.
Coated PTFE	Carbon nanotubes	750	50	20	2.8	[42]
Mixed matrix PVDF	Carbon black	675	20	24	2.3	This work

The two membranes were tested under similar conditions in terms of vacuum pressure, feed temperature and light irradiance and showed comparable distillate fluxes. These results confirmed the potential of mixed matrix membranes for DSMD application. The advantages of preparing a single layer membrane are, primarily, the easier procedure required and the more robust final structure. On the contrary, when a commercial membrane is coated with photothermal conversion materials, a poor adhesion between the support material and the surface skin can induce the delamination phenomena that can undermine the integrity and the performance of the membrane.

4. Conclusions

A simple and innovative preparation method to produce photoactive membranes was proposed, which involves incorporating the light absorbing filler directly inside the membrane matrix. The preparation procedure was based on a traditional NIPS technique and the filler was first dispersed in the polymeric dope solution. In line with the goals of an environmentally sustainable and safe membrane preparation process, a green solvent, triethyl phosphate, was used. The main novelty in this work was incorporating the selected filler, carbon black, inside the entire polymeric matrix. This particular formulation allowed to produce favorable outcomes on the internal membrane structure. Nearly all of the strategies to achieve surface heating MD are currently based on membrane surface modification by various techniques. However, based on our findings, inclusion of carbon black nanoparticles inside the whole membrane structure seems to have multiple beneficial effects on the membrane performance. In fact, the filler acts not only on the photothermal properties enhancing the light absorbance, but also on the membrane morphology increasing porosity and pores size. These latter characteristics, which are very important to reduce the membrane mass transfer resistance, were improved with the rising CB content.

In terms of photothermal properties, membranes prepared with a sufficient CB loading were able to reach satisfactory surface temperatures when exposed to light sources.

On the whole, as the carbon black loading was raised, the effectiveness during direct solar membrane distillation operation was improved and the membranes prepared with the highest CB concentration showed the best flux enhancement factor.

Author Contributions: Conceptualization, A.C. and M.P.; methodology, M.P.; investigation, M.A., R.F., E.A.M. and M.P.; writing—original draft preparation, M.P. and C.C.; writing—review and editing, A.C. and C.C.; visualization, M.P.; supervision, A.C. and C.C. All authors have read and agreed to the published version of the manuscript.

Funding: This research received no external funding.

Conflicts of Interest: The authors declare no conflict of interest.

References

1. Lawson, K.W.; Lloyd, D.R. Membrane distillation. *J. Membr. Sci.* **1997**, *124*, 1–25. [CrossRef]
2. Khayet, M.; Matsuura, T. *Membrane Distillation: Principles and Applications*; Elsevier: Amsterdam, The Netherlands, 2011; ISBN 9780444531261.
3. Schofield, R.W.W.; Fane, A.G.G.; Fell, C.J.D.J.D. Heat and mass transfer in membrane distillation. *J. Membr. Sci.* **1987**, *33*, 299–313. [CrossRef]
4. Alkhudhiri, A.; Darwish, N.; Hilal, N. Membrane distillation: A comprehensive review. *Desalination* **2012**, *287*, 2–18. [CrossRef]
5. Rezaei, M.; Warsinger, D.M.; Lienhard, V.J.H.; Duke, M.C.; Matsuura, T.; Samhaber, W.M. Wetting phenomena in membrane distillation: Mechanisms, reversal, and prevention. *Water Res.* **2018**, *139*, 329–352. [CrossRef]
6. Eykens, L.; De Sitter, K.; Dotremont, C.; Pinoy, L.; Van Der Bruggen, B. How to Optimize the Membrane Properties for Membrane Distillation: A Review. *Ind. Eng. Chem. Res.* **2016**, *55*, 9333–9343. [CrossRef]
7. Schneider, K.; Hölz, W.; Wollbeck, R.; Ripperger, S. Membranes and modules for transmembrane distillation. *J. Membr. Sci.* **1988**, *39*, 25–42. [CrossRef]
8. Pagliero, M.; Khayet, M.; García-Payo, C.; García-Fernández, L. Hollow fibre polymeric membranes for desalination by membrane distillation technology: A review of different morphological structures and key strategic improvements. *Desalination* **2021**, *516*, 115235. [CrossRef]
9. Li, B.; Sirkar, K.K.; York, O.H. Novel Membrane and Device for Direct Contact Membrane Distillation-Based Desalination Process. *Ind. Eng. Chem. Res.* **2004**, *43*, 5300–5309. [CrossRef]
10. Xu, Y.; Zhu, B.K.; Xu, Y.Y. Pilot test of vacuum membrane distillation for seawater desalination on a ship. *Desalination* **2006**, *189*, 165–169. [CrossRef]
11. Matsuyama, H.; Berghmans, S.; Lloyd, D.R. Formation of hydrophilic microporous membranes via thermally induced phase separation. *J. Membr. Sci.* **1998**, *142*, 213–224. [CrossRef]
12. Eykens, L.; De Sitter, K.; Dotremont, C.; Pinoy, L.; Van der Bruggen, B. Characterization and performance evaluation of commercially available hydrophobic membranes for direct contact membrane distillation. *Desalination* **2016**, *392*, 63–73. [CrossRef]
13. Ghani, F.A.; Hamzah, K.; Norharyati, W.; Salleh, W.; Mohamed, H. Preparation and characterization of PTFE flat sheet membrane: Effect of sodium benzoate content. *Malays. J. Fundam. Appl. Sci.* **2017**, *13*, 598–601. [CrossRef]
14. Pagliero, M.; Bottino, A.; Comite, A.; Costa, C. Novel hydrophobic PVDF membranes prepared by nonsolvent induced phase separation for membrane distillation. *J. Membr. Sci.* **2020**, *596*, 117575. [CrossRef]
15. Marino, T.; Russo, F.; Figoli, A. The Formation of Polyvinylidene Fluoride Membranes with Tailored Properties via Vapour/Non-Solvent Induced Phase Separation. *Membranes* **2018**, *8*, 71. [CrossRef] [PubMed]
16. Khayet, M.; Feng, C.Y.; Khulbe, K.C.; Matsuura, T. Preparation and characterization of polyvinylidene fluoride hollow fiber membranes for ultrafiltration. *Polymer* **2002**, *43*, 3879–3890. [CrossRef]
17. Pagliero, M.; Comite, A.; Soda, O.; Costa, C. Effect of support on PVDF membranes for distillation process. *J. Membr. Sci.* **2021**, *635*, 119528. [CrossRef]
18. Hubadillah, S.K.; Othman, M.H.D.; Matsuura, T.; Rahman, M.A.; Jaafar, J.; Ismail, A.F.; Amin, S.Z.M. Green silica-based ceramic hollow fiber membrane for seawater desalination via direct contact membrane distillation. *Sep. Purif. Technol.* **2018**, *205*, 22–31. [CrossRef]
19. Pagliero, M.; Bottino, A.; Comite, A.; Costa, C. Silanization of tubular ceramic membranes for application in membrane distillation. *J. Membr. Sci.* **2020**, *601*, 117911. [CrossRef]
20. Fang, H.; Gao, J.F.; Wang, H.T.; Chen, C.S. Hydrophobic porous alumina hollow fiber for water desalination via membrane distillation process. *J. Membr. Sci.* **2012**, *403–404*, 41–46. [CrossRef]
21. Eykens, L.; De Sitter, K.; Dotremont, C.; Pinoy, L.; Van der Bruggen, B. Membrane synthesis for membrane distillation: A review. *Sep. Purif. Technol.* **2017**, *182*, 36–51. [CrossRef]
22. Comite, A.; Pagliero, M.; Costa, C. Wastewater treatment by membrane distillation. In *Current Trends and Future Developments on (Bio-) Membranes*; Elsevier: Amsterdam, The Netherlands, 2020; pp. 3–34, ISBN 9780128168240.
23. Drioli, E.; Ali, A.; Macedonio, F. Membrane distillation: Recent developments and perspectives. *Desalination* **2015**, *356*, 56–84. [CrossRef]
24. González, D.; Amigo, J.; Suárez, F. Membrane distillation: Perspectives for sustainable and improved desalination. *Renew. Sustain. Energy Rev.* **2017**, *80*, 238–259. [CrossRef]
25. Rodríguez-Maroto, J.M.; Martínez, L. Bulk and measured temperatures in direct contact membrane distillation. *J. Membr. Sci.* **2005**, *250*, 141–149. [CrossRef]
26. Tan, Y.Z.; Chandrakant, S.P.; Ang, J.S.T.; Wang, H.; Chew, J.W. Localized induction heating of metallic spacers for energy-efficient membrane distillation. *J. Membr. Sci.* **2020**, *606*, 118150. [CrossRef]
27. Anvari, A.; Kekre, K.M.; Azimi Yancheshme, A.; Yao, Y.; Ronen, A. Membrane distillation of high salinity water by induction heated thermally conducting membranes. *J. Membr. Sci.* **2019**, *589*, 117253. [CrossRef]
28. Said, I.A.; Wang, S.; Li, Q. Field Demonstration of a Nanophotonics-Enabled Solar Membrane Distillation Reactor for Desalination. *Ind. Eng. Chem. Res.* **2019**, *58*, 18829–18835. [CrossRef]
29. Bamasag, A.; Alqahtani, T.; Sinha, S.; Ghaffour, N.; Phelan, P. Experimental investigation of a solar-heated direct contact membrane distillation system using evacuated tube collectors. *Desalination* **2020**, *487*, 114497. [CrossRef]

30. Chandrashekara, M.; Yadav, A. Water desalination system using solar heat: A review. *Renew. Sustain. Energy Rev.* **2017**, *67*, 1308–1330. [CrossRef]
31. Zaragoza, G.; Andrés-Mañas, J.A.; Ruiz-Aguirre, A. Commercial scale membrane distillation for solar desalination. *NPJ Clean Water* **2018**, *1*, 1–6. [CrossRef]
32. Karanikola, V.; Moore, S.E.; Deshmukh, A.; Arnold, R.G.; Elimelech, M.; Sáez, A.E. Economic performance of membrane distillation configurations in optimal solar thermal desalination systems. *Desalination* **2019**, *472*, 114164. [CrossRef]
33. Huang, J.; Hu, Y.; Bai, Y.; He, Y.; Zhu, J. Novel solar membrane distillation enabled by a PDMS/CNT/PVDF membrane with localized heating. *Desalination* **2020**, *489*, 114529. [CrossRef]
34. Chen, Y.-R.; Xin, R.; Huang, X.; Zuo, K.; Tung, K.-L.; Li, Q. Wetting-resistant photothermal nanocomposite membranes for direct solar membrane distillation. *J. Membr. Sci.* **2021**, *620*, 118913. [CrossRef]
35. Wu, J.; Zodrow, K.R.; Szemraj, P.B.; Li, Q. Photothermal nanocomposite membranes for direct solar membrane distillation. *J. Mater. Chem. A* **2017**, *5*, 23712–23719. [CrossRef]
36. Dongare, P.D.; Alabastri, A.; Pedersen, S.; Zodrow, K.R.; Hogan, N.J.; Neumann, O.; Wu, J.; Wang, T.; Deshmukh, A.; Elimelech, M.; et al. Nanophotonics-enabled solar membrane distillation for off-grid water purification. *Proc. Natl. Acad. Sci. USA* **2017**, *114*, 6936–6941. [CrossRef]
37. Pagliero, M.; Comite, A.; Costa, C.; Rizzardi, I.; Soda, O. A Single Step Preparation of Photothermally Active Polyvinylidene Fluoride Membranes Using Triethyl Phosphate as a Green Solvent for Distillation Applications. *Membranes* **2021**, *11*, 896. [CrossRef] [PubMed]
38. Lázaro, M.J.; Calvillo, L.; Celorrio, V.; Pardo, J.I.; Perathoner, S.; Moliner, R. *Study and Application of Carbon Black Vulcan XC-72R in Polymeric Electrolyte Fuel Cells*; Nova Science Publishers, Inc.: New York, NY, USA, 2011.
39. Elvidge, C.D.; Keith, D.M.; Tuttle, B.T.; Baugh, K.E. Spectral identification of lighting type and character. *Sensors* **2010**, *10*, 3961–3988. [CrossRef]
40. Jessen, W.; Wilbert, S.; Gueymard, C.A.; Polo, J.; Bian, Z.; Driesse, A.; Habte, A.; Marzo, A.; Armstrong, P.R.; Vignola, F.; et al. Proposal and evaluation of subordinate standard solar irradiance spectra for applications in solar energy systems. *Sol. Energy* **2018**, *168*, 30–43. [CrossRef]
41. Wu, X.; Jiang, Q.; Ghim, D.; Singamaneni, S.; Jun, Y.-S. Localized heating with a photothermal polydopamine coating facilitates a novel membrane distillation process. *J. Mater. Chem. A* **2018**, *6*, 18799–18807. [CrossRef]
42. Ma, Q.; Xu, Z.; Wang, R.; Poredoš, P. Distributed vacuum membrane distillation driven by direct-solar heating at ultra-low temperature. *Energy* **2022**, *239*, 121891. [CrossRef]

Article

The Application of Open Capillary Modules for Sweeping Gas Membrane Distillation

Marek Gryta

Faculty of Chemical Technology and Engineering, West Pomeranian University of Technology in Szczecin, ul. Pułaskiego 10, 70-322 Szczecin, Poland; marek.gryta@zut.edu.pl

Abstract: The paper presents the sweeping gas membrane distillation realised by using the capillary module (length 1.1 m and area 0.1 m^2) without housing (module shell). During the tests, the feed was flowing inside the hydrophobic polypropylene membranes. The studies were performed for two variants of process: with pre-heating (313–330 K) and without heating of the feed (brines). Under low gas flow (0.005 m/s) the evaporation performance varied in the range of 0.15–0.25 L/m^{2}h, depending on the relative humidity (42–63%) and the air temperature (293–300 K). The application of feed pre-heating to 330 K led to an increase in the evaporation performance to 2.4 L/m^{2}h. The permeate flux increased by 60% when the air flow velocities between the capillaries increased to 1.8–2.5 m/s. Increasing the feed flow rate from 0.1 to 0.59 m/s led to increase the permeate flux about 20% for feed temperature 293–310 K, and over 55% for feed temperature higher than 323 K.

Keywords: sweeping gas membrane distillation; membrane evaporation; submerged module

Citation: Gryta, M. The Application of Open Capillary Modules for Sweeping Gas Membrane Distillation. *Energies* **2022**, *15*, 1454. https://doi.org/10.3390/en15041454

Academic Editor: Alessandra Criscuoli

Received: 14 January 2022
Accepted: 15 February 2022
Published: 16 February 2022

Publisher's Note: MDPI stays neutral with regard to jurisdictional claims in published maps and institutional affiliations.

1. Introduction

Membrane distillation (MD) is performed by phase change; therefore, this process is energy-consuming and over 2500 kJ/kg of permeate obtained is required [1]. The water released from the feed in the form of vapour flows to the other side of the membrane where it is condensed, e.g., in the cold distillate stream. Such a case is called direct contact MD (DCMD) [2,3]. In the MD process, porous membranes made of hydrophobic polymers with a low thermal conductivity coefficient are used [4–6]. However, due to the small thickness of the membranes (e.g., 100 μm), it does not eliminate the heat conduction from the feed to the permeate side and, as a result, heat losses in the DCMD process may exceed 50% [2,7]. This inconvenience was limited by separating the membrane from the cold distillate with a gas layer (Air Gap MD) [8]. In another variant of the MD process, a gas stream flows on the permeate side [1,5]. A flowing gas is used to sweep the vapour out of the membrane permeate side, and this variant is called sweeping gas MD (SGMD) [1,3–6]. The gaseous layer increases the resistance to heat transfer which decreases the heat loss by conduction. For this reason, the application of the SGMD variant allowed to reduce heat losses to the level of 20% [9]. However, it must be recognised that in order to obtain freshwater from salt water in the SGMD variant, an external vapour condenser should be additionally used [1,10].

Water can also be desalinated by evaporating it from a wet surface of hydrophilic membranes. However, such evaporation caused a rapid crystallization of salts on the membrane surface [11], hence, the hydrophilic membranes can be only used for the separation of feed without solutes. When the hydrophobic membranes are applied, as in the SGMD process, the evaporation of water proceeds at the feed/membrane interface and a cross-flow of non-saturated feed prevents the precipitation of solutes, even for a high concentrated brine [12].

The driving force for mass transport in the MD process is the difference in vapour pressure [2,9]. In the case of SGMD, it results from the vapour pressure at the evaporation surface in the pores of the membrane and the water vapour content in the gas [1,5]. Mass and heat transfer causes both the concentration and temperature in the membrane adjacent layers to differ from those in the bulk (polarization effects), which reduces the driving force [2,5,9]. It has been reported that increasing the turbulence flow of the stream usually allows to limit the influence of polarization phenomena [10,13,14]. The polarization is particularly high on the side of the gas, which quickly becomes saturated with water vapour [15]. However, in classic modules with high membrane packing density, increasing the gas flow velocity is limited due to a significant increase in flow resistance [4]. To solve this problem, in the present work, an idea of open capillary modules was applied. In the modules, the water evaporation takes place from the surface of capillary membranes, whose bundles are loosely distributed inside large chambers. In such a system, the use of even high gas flow velocities do not cause a significant pressure drop. Apart from the flow velocity, the efficiency of the SGMD process is also influenced by the temperature of the feed and gas [1,4]. Increasing the feed temperature causes an exponential increase in the saturated vapour pressure, which leads to a significant increase in the driving force of mass transport [15,16]. As a result, the efficiency of the MD process increases, regardless of its variant [17–19].

It must be stressed that, in the published articles, the presented assessment of the influence of the process parameters on the SGMD course differs many times. Indeed, increasing the feed flow velocity most often significantly increased the permeate flux [10,15,17,20]; however, in other studies, a slight influence of this parameter was shown [14,18]. The gas temperature at the inlet to the module usually has little effect on the process performance [4,15,21], although its importance was shown in [4]. The discrepancies in the presented results are generally due to the design of the membrane modules used and their size. For example, when the membranes surface is small, the evaporation of water does not cause a significant change in the temperature of the gas transported in large amounts, hence, the influence of the gas temperature on the SGMD efficiency can be observed [21,22]. In the case of using modules with a large membrane surface (e.g., 1–2 m^2), due to the mass and heat transport, the gas temperature quickly becomes close to the temperature of the feed and, as a result, this temperature affects the efficiency of the SGMD process [4,22].

The use of small SGMD modules for testing allows for very favourable yields, often above 20 L/m^{2}h [15,16,18,20]. Unfortunately, similarly to large DCMD modules, increasing the process scale causes a multiple decrease in the permeate flux [4–6,13,23]. The amount of water vapour that the gas can absorb is small and increases with temperature, e.g., for air from 12.5 to 40.8 g/m^3 with a temperature change from 293 to 313 K. This means that for a temperature of 313 K, the evaporation of 10 kg/h of water requires the supply of air to the module in the amount over 245 m^3 (0.068 m^3/s). In the tested industrial modules, due to high flow resistance, the gas flows used were much smaller and, as a result, the gas quickly reached the saturation state, which significantly reduces the SGMD efficiency. For instance, in [4], for the Celgard Liqui-Cel® Extra-Flow module with an area of 1.4 m^2, the permeate flux of 0.5 L/m^{2}h was obtained. The efficiency at a level of 0.1 L/m^{2}h was obtained for a similar large module in work [23]. In this case, this was due to the fact that the permeate flux was only calculated by measuring the difference in gas humidity at the inlet and outlet of the module. Meanwhile, during tests in the SGMD installation with a vapour condenser, it was repeatedly found that the amount of water sweep by the gas from the module was greater than that resulting from the change in gas humidity [5,17,22]. This was explained in [22,23], where it was shown that after the gas is saturated, water vapour condenses and is removed from the module in the form of a mist. These observations were also confirmed in tests with the use of the industrial module [4,6].

A large evaporation surface can be obtained by using capillary modules [1,4,23]. However, in this case it is difficult to maintain a uniform gas flow between the capillaries, especially for higher packing density of the membranes [10,24]. The conclusions from these works are similar to those obtained during tests of DCMD capillary modules, the efficiency of which increased significantly when the arrangement of capillaries ensured static mixing [25]. The best conditions for mass transport in the MD module can be obtained by using cross-flow of the feed and permeate streams [14], which was successfully applied in the construction of a small SGMD installation proposed for freshwater production in remote area [1]. However, it can be expected that in large modules, e.g., with a capillary length of 1 m, the cross-flow of gas stream will cause the "sail effect". This effect for higher gas flow velocities (e.g., 5 m/s) may, due to increased stresses, cause the capillaries to break at the point of their attachment to module head. For this reason, the design of SGMD modules should ensure low gas flow resistance for its high flow velocities [4]. Such a possibility is given by the proposed open SGMD module design, where, additionally, flexible mounting of the capillary bundles can be used, which should limit the influence of the "sail effect".

In the modules with shell, both the temperature of the streams and the feed concentration change along the module, which causes a decrease in the efficiency of the MD process [15]. Uniformity of parameters along the entire surface of the capillary membranes was obtained in the MD process using submerged DCMD modules [7]. Worthy of note, similar conditions can be achieved in the case of SGMD by using modules with membranes loosely placed inside large chambers with natural convection or forced air flow generated by fans. Due to the significant polarization on the gas side, the use of fans increasing the gas flow velocity should significantly increase the efficiency of the process [15]. In the SGMD process, the use of the feed flow between capillaries is also possible. Indeed, it was carried out in the tests of 30 cm long module [16]. However, in the case of longer modules, the pressure of the gas flowing inside capillaries will increase significantly, which may increase the costs of gas pumping [4] and its bubbling through the pores of the membrane to the feed [21]. Therefore, in the SGMD process, the feed flow is generally used inside the capillaries [1,4,10,23].

MD module efficiency is strongly affected by the capillaries distribution configurations and membrane packing density [10,24,25]. In the proposed open capillary modules, the packing density of the membranes can be reduced, e.g., for membranes with a diameter of 2.6 mm spaced every 1 cm it would be 52 m^2/m^3. This value is several times lower than the value of 293 m^2/m^3 in Celgard Liqui-Cel® Extra-Flow module, which did not allow to obtain good conditions for the SGMD process [4]. The advantage of reducing the packing density is facilitated gas flow. Moreover, in a hot, sunny remote area, a large chamber with membranes could also function as a heat exchanger for air heating, which would significantly simplify the construction of the installation. In the presented work, the effectiveness of the SGMD process implemented in a capillary module without an external shell was tested. The aim of the research was to determine the influence of the process parameters on the level of real efficiencies of SGMD; hence, a module with a length similar to industrial modules was applied. Additionally, it was assessed whether, instead of feed pre-heating inside an external heat exchanger, it is possible to evaporate the water using only energy taken from the air flowing in the module.

Theory

The pores of the hydrophobic membranes applied in the SGMD process are non-wetted, and the feed evaporates from the feed/gas interface created inside the pores. The driving force for the water evaporation is created by a difference between the partial pressure of water vapour (in equilibrium with liquid feed), and its value in the air surrounding the membranes (Figure 1). The obtained permeate flux is proportional to the driving force, which is usually expressed by the application of mass transfer coefficient (K_m) [17,22]:

$$J = K_m(P_F - P_{Air}) = K_m \Delta P \tag{1}$$

where P_F and P_{Air} correspond to the saturated vapour pressure above the evaporation surface and the vapour pressure in the gas (air) stream, respectively.

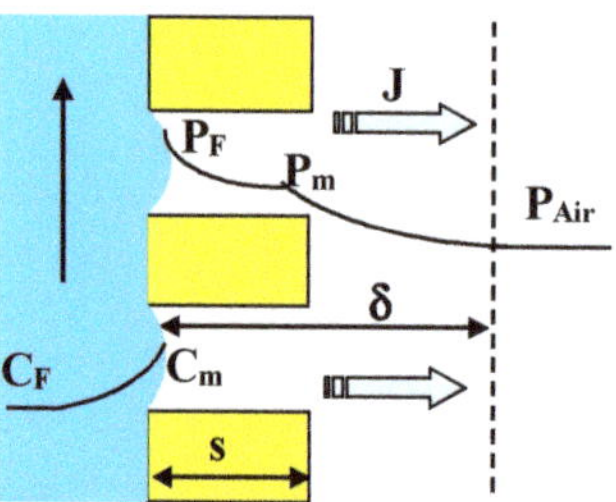

Figure 1. Water evaporation through non-wetted (hydrophobic) membranes. δ—thickness of the viscous boundary layer, s—membrane thickness, C_F—feed concentration, J—permeate flux.

The partial pressure of water vapour is strongly affected by the liquid temperature, and its value increases exponentially with increasing feed temperature, which can be expressed by the following equation [26]:

$$P_F(T) = \exp\left(8.07131 - \frac{1730.63}{233.42 + T}\right) \tag{2}$$

where the units of T and P_F are °C and mmHg, respectively.

During the streams flowing inside the SGMD module, the value of driving force may change [6,15]. The water evaporation increases the solutes concentration in the feed (Figure 1—C_m) which, in turn, decreases the P_F value. Moreover, the gas temperature and humidity fast increases, which strongly affects the SGMD module performance [4,26].

The vapour diffusion across the membrane pores creates the resistance for mas transfer [27]. The Knudsen or molecular diffusion mechanism influences, in a different degree, on this resistance due to a pore size distribution and the process conditions (e.g., temperature level). Taking these parameters into account, the K_m coefficient is expressed in the form [17,18,22]:

$$K_m = \frac{\varepsilon}{\chi s} \frac{M}{RT_m} D_{WA} \tag{3}$$

with the following parameters: porosity (ε), thickness (s), tortuosity (χ), molecular mass of water (M), gas constant (R), membrane temperature (T_m), and effective diffusion coefficient (D_{WA}). In the case of membrane with the pores below 0.1 μm the effective diffusion is dominated by the Knudsen diffusion [6,22].

In the case when the water is evaporated into the ambient air, a viscous boundary layer is formed above the evaporation surface through which the diffusion of water vapour takes places [10]. The rate of liquid volume change evaporated from the wet surface under the isothermal conditions can be determined using Fick's law [11]:

$$\frac{dV}{dt} = \frac{D\,A\,M}{\delta\,\rho\,R\,T}(P_F - P_{Air}) \tag{4}$$

where D is the diffusion coefficient of liquid particles in the air, A is the liquid surface area available for evaporation, δ is the thickness of viscous boundary layer, ρ is a density of liquid.

An important point which should be noted is that the thickness of the viscous boundary layer formed above the membrane surface can be reduced by increasing turbulence in the gas flow [15]. However, this method cannot change the value of membrane thickness (s), which contributes to the value of δ (Figure 1); hence, the thickness of the membrane has a significant effect on the SGMD process efficiency [14]. Moreover, the value of P_{Air} is associated with air humidity, expressed e.g., by relative humidity (RH), thus, water evaporation increasing a value of the humidity also decreases the evaporation rate. For this reason, the process efficiency is limited in the modules with shell, because a value of the relative humidity of the air in the SGMD module rapidly increases even for high flow rates of air [22,27].

The maximal content of water vapour in air at given temperature is expressed by following equation [10,22]:

$$X = 0.622 \frac{P_S}{P_{atm} - P_S} \tag{5}$$

where P_S is water vapour pressure in saturated air and P_{atm} is atmospheric pressure.

Evaporation is an energy intensive process, therefore, the transport of heat to the interface is the rate-limiting step for the evaporation of liquids into an inert gas. Depending on the membrane installation design and process conditions, the temperature in the gas phase near the liquid–gas interface can be higher or lower than that of the liquid (Figure 2). The application of feed temperature higher than the air temperature allows to increase the driving force of process, hence, pre-heating of feed (Figure 2b) is usually realised in SGMD process [4,14,15,18]. In this case, the limitation is a small amount of water vapour that caused the saturation of the air, which can lead to unfavourable condensation of vapour inside the membrane module [22,26]. The vapour condensation on the membrane surface can be avoided when the modules without an external shell will be used, which also allows to curry out the evaporation without heating of the feed (natural evaporation). In this case, the bulk temperature of the air is higher than the temperature of the evaporating feed [4,15].

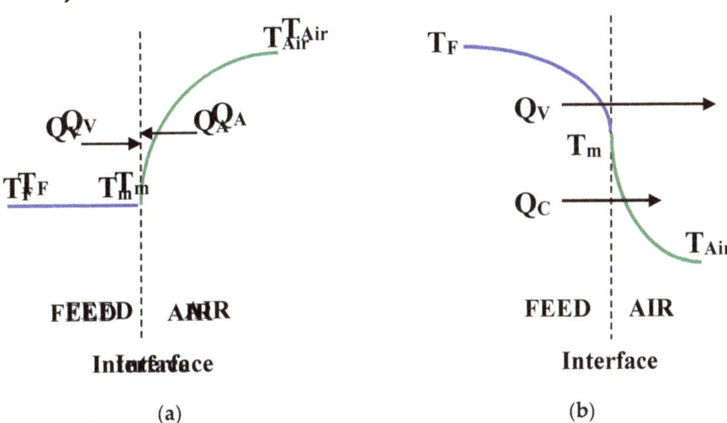

Figure 2. Temperature profile at feed–air interface Membrane evaporation of water: (**a**) without feed heating; (**b**) with feed pre-heating. Q—heat.

A temperature profile without feed heating is presented in Figure 2a. The energy for evaporation (Q_V) can only be taken from the gas phase (Q_A); therefore, the water temperature is quickly aligned to the constant value due to a larger thermal conductivity and the temperature gradient in liquid is negligible. The heat transfer is described by the following equations [9,28]:

$$Q_V = J \, \Delta H_V \tag{6}$$

$$Q_V = Q_A = h \, (T_{Air} - T_m) \tag{7}$$

where ΔH_V is the latent heat of water vapour evaporation and h is the convective heat transfer coefficient (on the air side). When the feed is heated, a part of the feed energy is lost by conductivity (Figure 2b, Q_C) into the air [26]:

$$Q_C = H \, (T_F - T_{Air}) \tag{8}$$

where H is the overall heat transfer coefficient.

The thermal efficiency (E) of MD process can be determined from the following relationship [9]:

$$E = Q_V / (Q_V + Q_C)$$ (9)

The value of convective heat transfer coefficient can be calculated from the Nusselt number estimated from following correlation [28]:

$$Nu = \frac{h\,d_h}{\lambda} = 4.36 + \frac{0.036 Pe\frac{d_h}{L}}{1 + 0.0011\left(Pe\frac{d_h}{L}\right)^{0.8}}$$ (10)

where Nu is Nusselt number, Pe is the Peclet number, L—length of channel (e.g., membrane capillary), d_h—hydraulic diameter, λ—heat conductivity coefficient. The remaining model equations applied for calculation of submerged MD modules were presented in work [7].

2. Materials and Methods

The studies of open capillary modules applied for SGMD were carried out using an installation presented in Figure 3. During experiments with feed pre-heating the feed tank was immersed inside a water thermostat, which allowed to control the feed temperature.

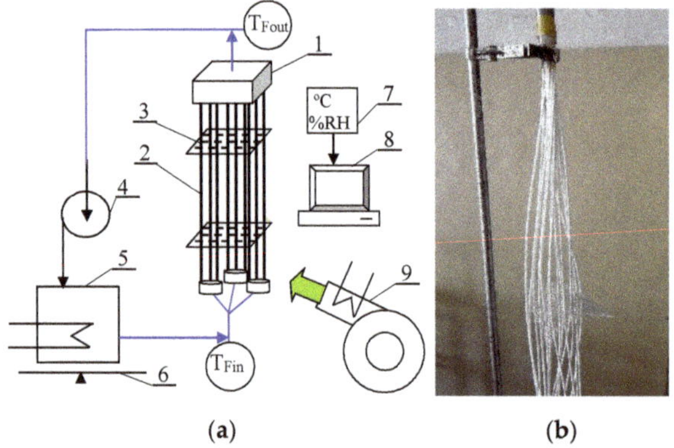

Figure 3. Experimental set-up. (**a**) Diagram: 1—SGMD membrane module, 2—capillary membrane, 3—polypropylene net, 4—peristaltic pump, 5—feed tank (thermostatic), 6—balance, 7—hygrometer, 8—computer, 9—fan with heat element, T$_{Fin}$, T$_{Fout}$—thermometers. (**b**) Photo of the open module.

The commercial hydrophobic K1800 polypropylene capillary membranes, manufactured for microfiltration (Euro-Sep, Warszawa, Poland), were used for the studies of membrane evaporation. A module was equipped with 16 capillary membranes, which were glued on both ends inside the PCV tube (1/2″ diameter). The capillary membranes have the internal diameter of 1.8 mm and outer diameter of 2.6 mm, and the effective length of 1.1 m. The total membranes area calculated for the lumen side amounted to 0.1 m². The membranes were positioned rectangular in every third mesh of two polypropylene nets. A distance between each capillary membrane was about 1 cm. The obtained module packing fractions was 0.05 and membrane packing density 52 m²/m³.

In [1], a horizontal arrangement of capillary membranes was proposed, which is possible in the case of short capillaries. Filling long horizontal capillaries with water would cause their significant deflection and stress in the places of their attachment, which would increase the risk of capillary breakage. For this reason, in the modules used, the capillaries were mounted vertically, with flexible mounting of the lower head of the module, as schematically shown in Figure 3. This solution enables a capillary bundle wave, which allows eliminating the negative influence of the "sail effect", which may be important for higher gas flow velocities through the chamber.

In this work, most of the tests were carried out in a room with air circulation caused by the ventilation system, which, for the required multiple air changes per hour, assured the air flow in the range of 0.002–0.005 m/s. Tests were also carried out with air forced by the fan from the bottom to the top of the capillaries. In this case, the gas flow velocity in the lower part of the module was 2.3–2.5 m/s and along the module it decreased to 1.8 m/s in the upper part. The fan did not have rotor speed control, hence, the obtained air flows resulted from the factory efficiency of the fan. Whereas the fan was equipped with an electrical heat element, which made it possible to additionally carry out tests with hot air (313 K). The air flow velocity was measured using electronic anemometer MT-881 (MeasureMe, China).

The feed flowed inside the capillaries (lumen side) during the evaporation experiments. A peristaltic pump was used, and the feed flow rate was equal to 0.1 m/s. The influence of flow velocity (0.1–0.59 m/s) on the evaporation efficiency was additionally investigated.

The total dissolved solids (TDS) of solutions were measured with a 6P Ultrameter (Myron L Company, Carlsbad, CA, USA). This meter was calibrated for measurements as NaCl using TDS/conductivity standard solution (Myron L Company). The air temperature and relative humidity were measured by electronic hygrometer AZ8829 (AZ-Instruments, Kraków, Poland) connected with computer software TRLOG v. 3.4. The feed temperature was measured using electronic thermometers PT-401 with measurement accuracy 0.1 K (Elmetron, Zabrze, Poland).

The membrane evaporation tests were carried out using distilled water or NaCl solutions (pure NaCl, Chempur, Piekary Śląskie, Poland) as a feed. The studies was conducted continuously for several months. Indeed, the experiments started in June (summer) and ended in October (autumn). The changes of module efficiency (maximum permeate flux) were measured periodically at established periods using distilled water as a feed. The MD installation was working continuously. The permeate flux was calculated every 20–24 h, based on the decrease of the feed volume.

3. Results

3.1. Influence of Process Parameters on the Permeate Flux

The presented work considers the possibility of implementing the SGMD process in open modules, which are created by bundles of capillaries distributed symmetrically in gas-filled chambers. Open SGMD module can operate in conditions similar to natural evaporation or in chambers with forced gas flow. In the first stage of studies, the operation of the open SGMD module under very low air flow conditions was tested.

The feed temperature is one of the most important parameters in the MD process. The performed tests confirmed that also in the case of open modules, the feed temperature has a significant impact on the SGMD process performance. The results presented in Figure 4 show that increasing the inlet feed temperature from 293 to 323 K resulted in a three–four-fold increase in the permeate flux, which, for 323 K, was equal to 2.4 L/m^{2}h (v_F = 0.59 m/s). It is essential to mention that similarly high increases in the process performance were obtained by increasing the feed temperature in traditional SGMD modules [5,17,20].

Water evaporation from the feed causes a significant increase in the concentration of solutes in the boundary layer (Figure 1—C_m). The results obtained during the separation of the solution containing 100–120 g NaCl/L are additionally presented in Figure 4. The noted permeate fluxes were only slightly lower than those obtained for distilled water. Therefore, it can be concluded that such a significant increase of concentration caused only a small decline of the performance, which is a known advantage of the MD process [7,12]. As a result, the MD process can be used not only for the preparation of freshwater from concentrated brines, but also for the concentration of solutions [14,18,29].

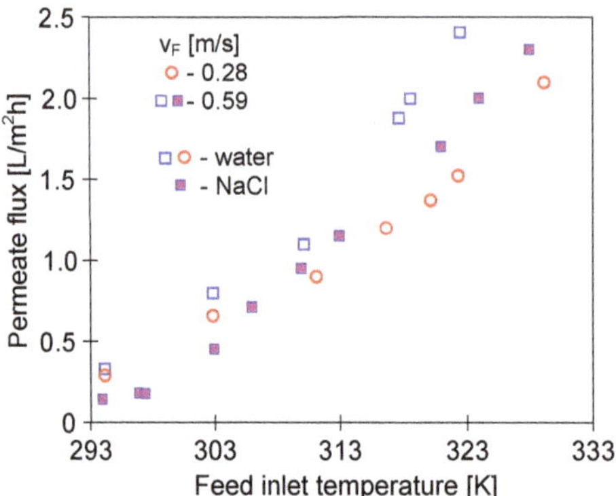

Figure 4. The influence of feed temperature and velocity on the permeate flux. Feed: distilled water and NaCl solutions (100–120 g/dm^3).

In the MD process, solutions can be concentrated up to the saturation state [12]. According to Raoult's law the vapour pressure decreases with increasing of feed water salinity as follows [30]:

$$P_F = (1 - x)\, P_F^0 \tag{11}$$

where P_F^0 is the vapour pressure of pure water and x is the molar fraction of salt in the water. For example, for almost saturated NaCl solution (5.5 mole/L) with a water concentration of 55.5 mole/L, we have x = 5.5/(55.5 + 5.5) = 0.09, which gives about a 10% decrease in driving force. The decrease in permeate flux obtained with saturated solutions was greater than 20%, which, in addition to reducing water activity, was also influenced by an increase in feed viscosity [31]. It is important to note that, during the MD of concentrated solutions, the feed flow conditions should be ensured to prevent salt crystallization on the membrane surface [12].

In addition, the obtained results (Figure 4) showed a significant influence of the feed flow velocity on the permeate flux. An important point which should be noted is that this effect increased with increasing feed temperature. Indeed, for instance, for T_F = 323 K, increasing v_F from 0.28 to 0.59 m/s resulted in an increase of the permeate flux by 55% (from 1.55 to 2.4 L/m^{2}h). During the feed flow through the module, due to mass and heat transfer, the feed temperature decreases, which can be reduced by increasing its flow velocity [7,20]. Conducting the SGMD process in small modules and at relatively high flows does not cause significant changes in the feed temperature [26]. Hence, in some studies a slight influence of the feed flow velocity was shown [26,32], which, however, changes when modules with a much larger membrane area are tested [10,17,20].

As stated before, in the presented work, the active length of the membranes in the capillary module was 1.1 m, which allowed to obtain a significant difference in the feed temperature between the inlet and outlet from the module (Figure 5). Similar changes in the feed temperature was presented in [20].

The obtained T_{Finlet}-$T_{Foutlet}$ values increased exponentially with the increase of the T_{Finlet} temperature as the amount of evaporated water also similarly increased (Figure 4). Increasing the feed velocity limited its temperature decline, e.g., from 12 to 4 K when the flow velocity has increased from 0.28 to 0.59 m/s (Figure 5). It should be noted that this stabilizing effect of the feed temperature results not only from the feed velocity but also from the channel cross-sectional area, which determines the ratio of the volumetric feed velocity (L/s) to the membrane area [m^2]. In the case of the tested capillary module, it was equal to 0.11 L/s m^2 (v_F = 0.28 m/s) and 0.24 L/s m^2 for the flow velocity of 0.59 m/s. For an example of a plate module with a size of 10 × 10 cm and a channel height of 0.2 cm, at feed velocity of 0.59 m/s, the value of this ratio is equal to 11.8 L/s m^2. Such a high value explains the slight changes in the feed temperature when testing such small modules. However, if the length of this plate module will be increased to 1 m, the obtained value will be equal to 0.118 L/s m^2. It will result in a significant decline in the feed temperature, which reduces the SGMD efficiency several times, e.g., to the level of 0.3 kg/m^{2}h [15].

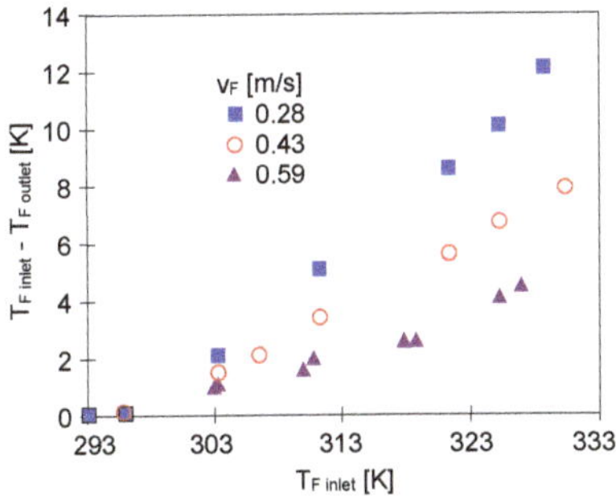

Figure 5. The influence of feed velocity and feed inlet temperature on the feed temperature decline.

It is necessary to mention that the feed flow velocity also has a significant impact on the value of the convective heat transfer coefficient (h), which determines the value of the temperature T_m (Figure 2). The temperature of the evaporation surface determines the vapour pressure at the interface phases (P_F—Figure 1, and Equation (2)). For the capillary membranes used, the value of the h coefficient can be calculated from the correlation expressed by Equation (10). For T_F = 323 K, the obtained h value was 2380 W/m^{2}K (v_F = 0.28 m/s) and 2465 W/m^{2}K for v_F = 0.59 m/s. The temperature distribution along the module calculated for the v_F = 0.59 m/s is presented in Figure 6. The difference in temperature T_{Fout} and T_m did not exceed 1 K, which corresponds to the values of the feed temperature profile presented in work [15].

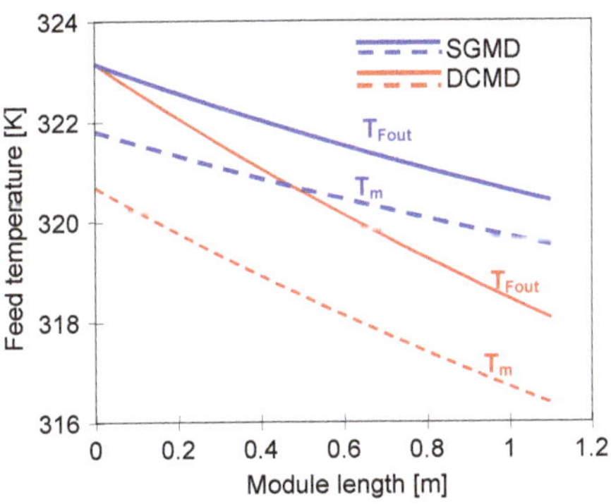

Figure 6. The distribution of feed temperatures profile (outlet and at the evaporation surface) calculated for SGMD and DCMD capillary modules.

The results of numerical calculations performed for a submerged DCMD module (similar to SGMD module) immersed in the distillate are additionally shown in Figure 6. All parameters from the SGMD module calculations were adopted, however, the air was replaced with distilled water at 293 K. It has been found that the value of the temperature polarization on the feed side is twice as large, which is due to the fact that in the DCMD variant, the membrane is in contact with the cold distillate, and this significantly increases the value of the conducted heat (Figure 2b, Q_C). As a result, the calculated thermal efficiency of the tested SGMD module, amounting to 68%, decreased to 43% in the case of its operation in the DCMD variant. Similar values of thermal efficiency for DCMD submerged modules were obtained in work [7].

Since the feed temperature decreases as it flows along the module, the module size has the significant impact on the MD process run. Indeed, for small modules, the T_F change is insignificant, hence the conversion fluxes (L/m^2h) are overestimated and many times greater than those obtained in modules whose area is actually 1 m^2 or more [13].

Effect of Gas Flow Rate

The thermal equilibrium between the air and the feed depends not only on the air temperature but also on the evaporation rate which is influenced by the boundary layer conditions (Equation (4)). For this reason, increasing the flow velocity, reducing the thickness of the layer δ [15], has a significant impact on the course of the SGMD process [23]. The reduction of the water vapor concentration at the membrane surface caused by the gas flow increases the value of the vapor pressure difference (dP/dx), which is schematically presented in Figure 7.

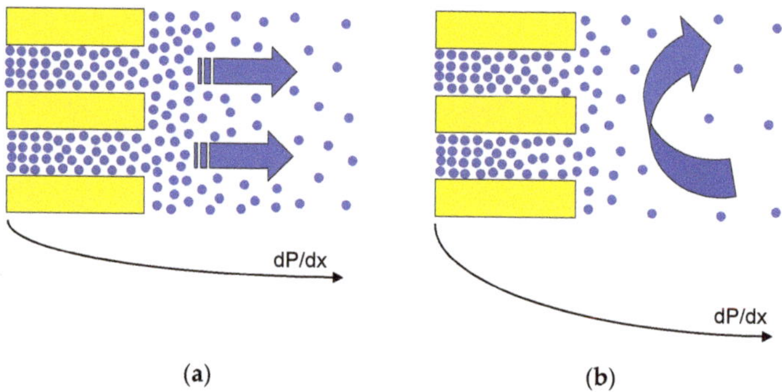

Figure 7. The changes of water vapour concentration in the boundary layer. (**a**) natural convection, (**b**) forced convection.

Forced convection of gas increases the rate of evaporation [4], which was confirmed by the results presented in Figure 8 (Tests 1 and 2—without feed pre-heating), showing that the permeate flux increased from 0.240 to 0.385 L/m^2h (60%) when the fan was started and air was flowing between the capillaries at a velocity of 1.8–2.5 m/s. A similarly large effect of increasing the gas flow velocity was demonstrated in SGMD modules with a shell [20,32]. However, the process efficiencies obtained for such classical modules are presented in a wide range of 0.1–50 L/m^2h [4,16,22]. Such different efficiencies result from the fact that the vapour condensation (mist) in the sweep gas significantly increases the efficiency of the tested modules [22,26]. In the studied case (Figure 8), there was no vapour condensation, hence, the use of an open structure made it possible to keep the vapour content in the air below its saturation value (Figure 9). Moreover, the performed measurements showed that the use of the fan reduced the air humidity between the capillaries. The air temperature inside the module also slightly decreased, which confirms that the energy used for the increased evaporation came from the air surrounding the membranes. The higher the air temperature, the more energy is transferred to the evaporation surface, which allowed to significantly increase the efficiency (Figure 8, Tests 3 and 4). The possibility of feed heating by the air surrounding the membranes is presented in the next Section.

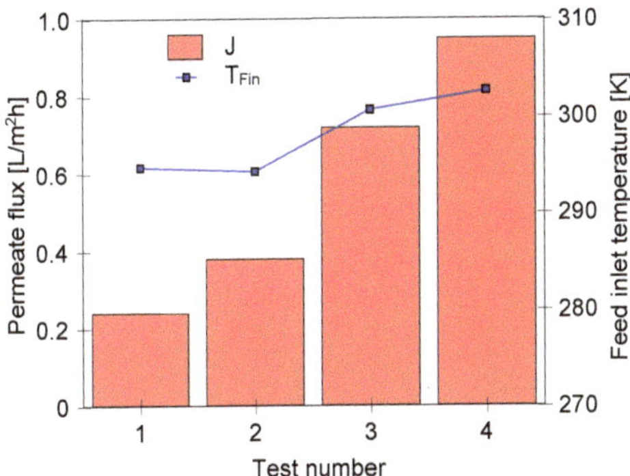

Figure 8. The comparison of SGMD performance obtained for natural convection (Test 1) and forced convection with air flow 1.8–2.5 m/s (Tests 2–4). Conditions: (1) RH = 40% for T_{Air} = 295 K, (2) RH = 47% for T_{Air} = 294.8 K, (3) RH = 39.3% for T_{Air} = 293.8 K, air heating: inlet 313 K and outlet 305 K, (4) RH = 44.9% for T_{Air} = 294.5 K, air heating: inlet 313 K and outlet 307 K.

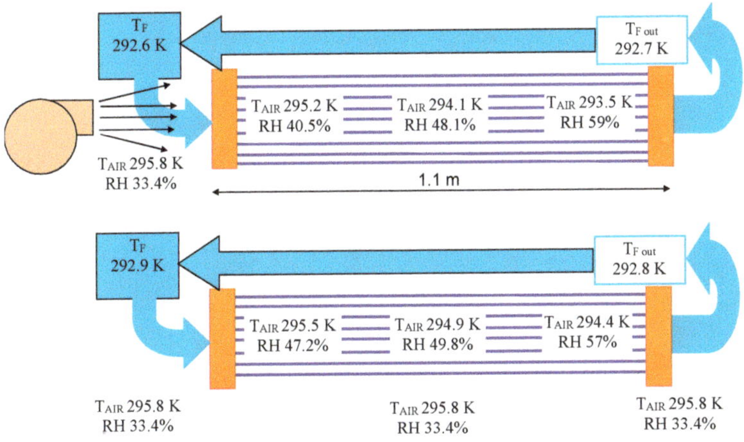

Figure 9. The changes of air temperature and humidity inside the capillary bundle during SGMD carried out with and without fan working. Parameters of air surrounding the module: T_{AIR} = 295.8 K and relative humidity 33.4%.

In the SGMD process assumption, the air flowing out of the module should not be saturated, which prevents vapour condensation in the module channels [5]. Notwithstanding, in the SGMD module, conditions are often created for supersaturation of the gas, which causes the formation of water droplets on the permeate side [22]. In this case, a significant part of the water evaporated from the feed is discharged from the module in the form of a mist, which significantly increases the calculated permeate flux [5,17,22,26]. The maximum amount of water vapour that the air can contain at a given temperature was calculated using Equations (2) and (9) and is shown in Figure 10. The amount above 0.1 kg H_2O/kg air increases rapidly, similar as does the water vapour pressure, for temperatures above 323 K. The obtained values were converted into the maximum flux that would result in air saturation at a given temperature. The results of the calculations performed show that, e.g., for a permeate flux of 30 L/m²h, the air supply (343 K) should be 90 m³/h per 1 m² of membranes. For the example flat channel dimension, 1 m wide and 5 mm high; this would correspond to a flow velocity of 5 m/s. Reducing the flow velocity to 1 m/s would allow about 5 kg/h of evaporating water to be removed from 1 m² of membranes. It follows that obtaining higher efficiencies in the SGMD process requires the use of high gas flow velocities, which is difficult to obtain in large modules due to the arising significant air flow resistance [4]. The proposed open modules, due to the use of a lower packing factor of the membranes, would facilitate the use of higher air flow velocities. For the tested module with air flow along the capillaries (cross-section 5 × 5 cm) with a speed of 2 m/s, an air flow of 18 m^3/h was obtained, with no visible effects of waving membranes.

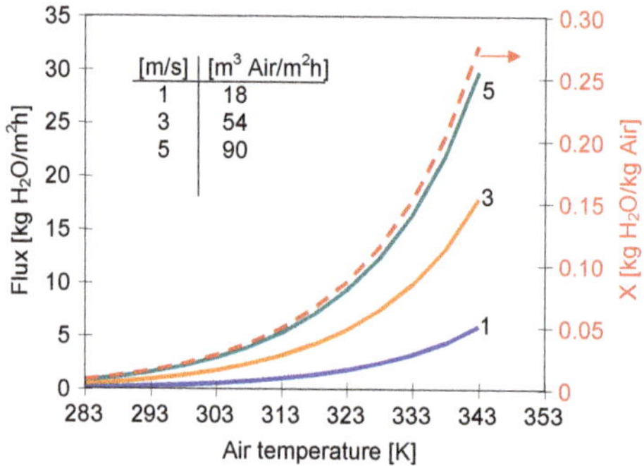

Figure 10. The influence of air temperature on the water vapour capacity (X) and maximum permeate flux possible for exemplary volume flow air velocity (m^3/m^{2}h) without vapour condensation inside SGMD module. Assumed air density 1.2 kg/m^3. Flow velocity [m/s] recalculated for channel cross section dimension 1 m × 0.5 cm.

The results shown in Figure 8 demonstrated that besides the gas flow velocity, the feed temperature also has a great influence on the SGMD performance. Increasing its value from 294 to 302 K resulted in an increase in the permeate flux from 0.38 to 0.95 L/m^{2}h. If temperature $T_F < T_{Air}$, the gas will be the energy source for water evaporation.

3.2. Feed Heating by Sweeping Gas

The heat of water vaporization is high (2500 kJ/kg), while the specific heat of air is low (1.005 kJ/kg), hence, its use as a heating medium requires forcing large volumes of gas through the module. In classic modules with a shell, which are characterised by a high degree of membrane packing density, it is not possible to obtain large gas flows due to a significant increase in flow resistance [4]. Open capillary modules allow for the implementation of SGMD in conditions of a significant excess of flowing gas in relation to the surface of the membranes, which makes it possible to use air to heat the feed. This variant is schematically shown in Figure 2a. The results presented in Figures 8 and 9 show that energy transfer from air to the membrane can be accomplished for both natural and forced convection.

In the MD process, the permeate volume obtained in relation to the feed is small, therefore, in MD installations multiple feed recirculation is used [33]. In the case when the external heat exchanger (feed pre-heating) is not used, the water evaporation in the module cool-down feed, but simultaneously the air surrounding the installation heats the feed. As a result of multiple recirculation, an equilibrium temperature is achieved, and the T_{Finlet} was closed to the T_{Fout} temperature (Figure 11). A variant of operation without feed pre-heating was used during summer studies, e.g., obtaining the feed temperature over 300 K on hot days (T_{Air} = 302 K), whereas the feed temperature along the module did not undergo changes. The tests were carried out without a fan working and the permeate flux was in the range of 0.2–0.3 L/m^{2}h. This noteworthy result indicates that the process performance depends not only on the temperature but also on the humidity of the air surrounding the installation, which changed during test in the range 49–57%.

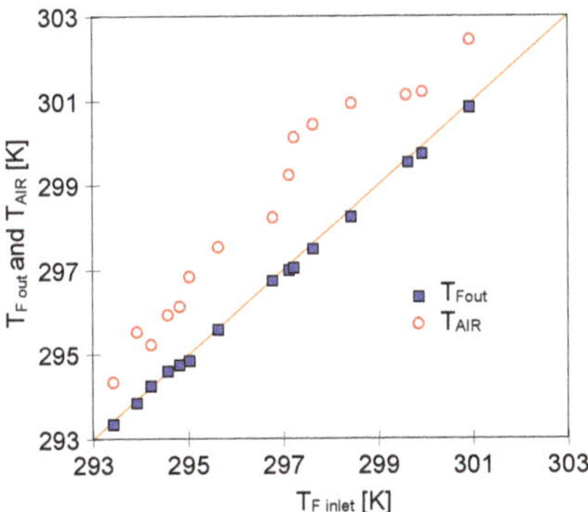

Figure 11. The influence of air and feed inlet temperature on the value of feed temperature at the module outlet (T_{Fout}). Feed velocity 0.1 m/s.

The permeate flux obtained for almost a constant feed temperature allows to determine the value of K_m coefficient from Equations (1) and (3). This part of studies was carried out using distilled water as a feed, and the value of $K_m = (1.3 +/- 0.15) \times 10^{-4}$ L/m²h Pa was obtained. The air parameters changes in the range: T_{Air} = 299–300.8 K and RH = 45–53% during this study. More than 10 times lower values ($1 \div 8 \times 10^{-5}$ L/m²h Pa) were obtained for industrial modules in [5]. This value increased with the increase of the gas flow velocity, which confirms the conclusion from Figure 10 that the gas is saturated quickly in large modules with shell [4,23]. As a result, the main component of the mass transport resistance is on the gas side, in contrast to the tested open module, where the parameters of the membranes determine the evaporation rate.

The results presented in Figure 8 indicated that the increase in SGMD performance can be obtained by increasing the gas flow velocity between the capillary membranes. Additionally, for the variant with forced gas flow, it is possible to heat the feed with the hot gas, which allows to increase the feed temperature and increase the performance of the process several times. In the tested case, the temperature of the air flowing along the membranes varied from 313 K (module bottom) to 299 K (the top of the module). As a result, the permeate flux was equal to 0.72 L/m²h when the feed temperature increased to 300.6 K and 0.95 L/m²h for T_F = 302.6 K (Figure 8). For comparison, the temperature distribution along the module for the case without air heating was also shown in Figure 9. In this case, the feed temperature will be lower than the air temperature, it tends to the wet bulb temperature, which lowers the vapour pressure in the pores of the membrane and, as a result, the evaporation efficiency is lower. In the case of small laboratory installations, it should be taken into account that not only the air parameters (T_{Air}, RH) but also the installation parameters (e.g., feed volume, tank, and tubing wall surface) affect the test result. The influence of the ratio of the feed volume to the membrane area on the stabilization of the installation operation is shown in Figure 12. It is worth noting that, in the case of starting large industrial modules, the SGMD stabilization period was about 2 h [23].

Heating the feed with the hot gas is a method that provides an to increase energy efficiency—as, once equilibrated, all heat is used for evaporation (Figure 2a). In contrast, when the module is fed with hot water, only part of the energy from the feed goes to evaporation (Figure 2b). In the case of the tested K1800 capillary membranes, a module with an area of 1 m² for v_F = 1.22 m/s supplies 0.5 kg/s of the feed, i.e., for T_F = 333 K, the feed energy is equal to 126 kW. Assuming that the permeate flux is 10 kg/m²h [7], only 6.9 kW is used for evaporation (ΔH = 2500 kJ/kg). The temperature at the outlet from the module for these conditions is about 10 K lower, which ensures a decrease in energy by 21 kW, i.e., a thermal efficiency of 33%.

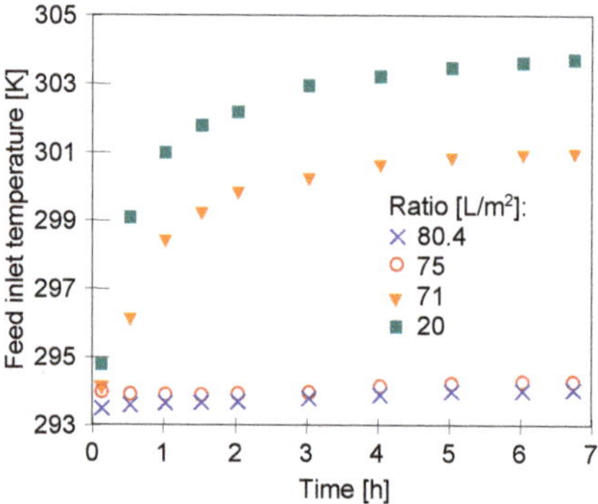

Figure 12. The changes of $T_{F\,inlet}$ during feed recirculation for different ratio of initial feed volume to membrane area. SGMD conditions correspond to test number from Figure 8: 80.4—1, 75—2, 71—3 and 20—4, respectively.

In SGMD modules, the separation of the membrane by a gas layer limited heat loss and the thermal efficiency increased to 50–75% [4]. This value depends mainly on the temperature of the feed. Considering only the drop in temperature of the feed flowing through the Celgard Liqui-Cel module (1.4 m^2) [4], the energy consumption for water evaporation were calculated and presented in Figure 13. As the temperature of the feed increases, the efficiency increases, but also the energy consumption approaches the minimum value of 694 kWh/m^3 (calculated for $\Delta H = 2500$ kJ/kg).

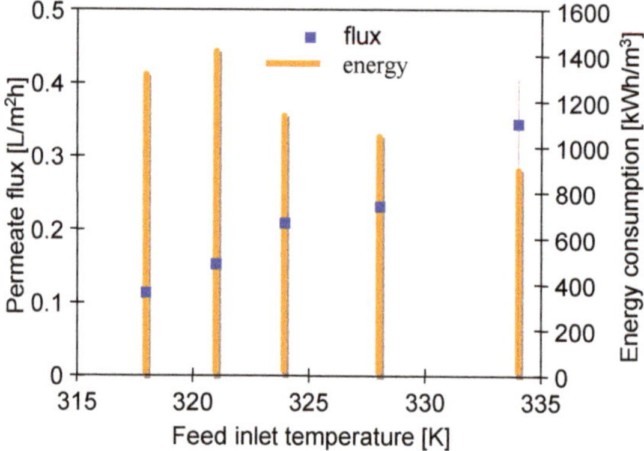

Figure 13. The influence of feed temperature on the permeate flux and energy amount consumed for water evaporation. Data for calculations taken from [4].

In the SGMD variant with the feed pre-heating (Figure 13), most of the energy consumed resulted from the drop in the temperature of the feed in the module. In the proposed open-module solution, energy for water evaporation is taken from the environment around the installation. The air heated by the sun flows between the capillaries and transfers its energy to the surface of the membranes. In this case, the energy consumption in the process will be mainly due to the operation of the fans. However, even large fans with a capacity of several thousand m^3/h consume much less energy compared to the heat of water vaporization [34].

The motor power of the fan used in addition to capacity is also influenced by the flow resistance of air inside the chamber. Importantly, the determination of their values for the air flow in the open module requires further research. However, considering that the flow resistance through the ventilation grille is in the order of 20–40 Pa, it can be assumed that the flow resistance through a chamber with a cross section of 0.5×1 m in which 5 rows of capillaries are arranged, each of which is located 1–2 cm from the other, should not exceed 500 Pa. Sample fan GMT-R-60 series equipped with a 340 W motor and capacity decreasing from 540 to 400 m^3/h with increasing flow resistance from 230 to 1000 Pa [34] should be appropriate for the open module in which 250 capillaries with a length of 1 m were installed (area of 1.4 m^2). This installation allows for the evaporation of a similar amount of water (Figure 8) as obtained in the Celgard Liqui-Cel module [4], which, however, consumed many times more energy (Figure 13).

3.3. Long-Term Studies

The MD process is most often proposed for water desalination, but other applications such as concentration of solutions are also contemplated. One of the examples is the use of SGMD to concentrate solutions of glycerol [29], sugar [32], or fruit juices [18]. In the case of solution concentration, there is no need to use a vapour condenser, which significantly simplifies the installation and allows the use of the open SGMD variant of the modules without the need to achieve a state of supersaturation of the vapour in the sweeping gas (inside condenser). The effectiveness of water evaporation in open capillary modules without the use of feed pre-heating has been proven in long-term studies.

The studies of SGMD were carried out for 4 months corresponding to over 1100 h of module work using distilled water with 5 g/L of NaCl as a feed. After such period of operation, the module performance was changed only slightly (Figure 14). Fluctuations of maximum permeate flux with time presented in this figure mainly resulted from the changes of ambient air temperature and relative humidity. The data presented in Figure 14 were calculated as the averages for each day. In order to achieve the operation conditions close to the natural ones (outdoor), the installation was left in a room with a slightly open window, which caused the changes of parameters of installation operation depending on weather conditions and time of day. The studies were started on summer (average air temperature 299 K) and were ended on autumn, when the average air temperature dropped to 295 K. The increase in humidity results not only from rainfalls, but also from increasing mean air temperature, which intensifies the water evaporation from soil and plants. It is worth noting that despite significant changes in air parameters, the obtained average daily performance was at a similar level. It also shows that the capillary membranes used were not wetted despite their long service life. It is well known that the membrane wetting is a major operational issue in the MD process. This problem also applies to the SGMD variant, as a result of which, the permeate flux decreased by 40% already after 140 h of the process run [16].

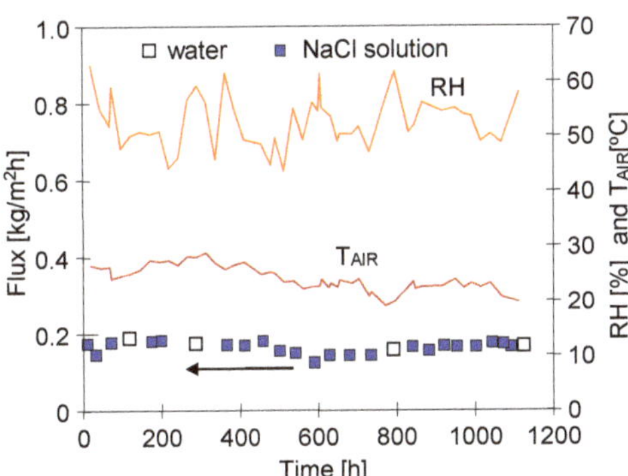

Figure 14. Changes the maximal permeate flux and average changes (daily) temperature and relative humidity of air surrounding the membrane module during studying of NaCl (5 g/L) solution evaporation.

To simplify the installation, the long-term tests were carried out without air flow forced by fan. However, the results presented in Figure 8 indicate that the use of fans would allow the permeate flux to be increased to a level of 0.3 L/m^{2}h. For the membrane packing density 52 m^2/m^3 used in the work, for an exemplary chamber 2 m wide and 5 m long, an installation with an area of over 500 m^2 would be obtained, which would allow for the evaporation of 150 L/h of water. The analysis of the impact of the packing degree on the gas velocity distribution presented in [24] shows that a two–three-fold increase in the packing density should not significantly affect the operation of the installation and the capacity of 500 L/h should be available for an example installation without feed pre-heating.

4. Conclusions

The performance obtained in the conducted research was similar to the SGMD results presented in the literature obtained for industrial membrane contactors, which indicates that the proposed open capillary modules are an interesting alternative to classic modules with shell.

The conducted tests confirmed that, in the case of industrial SGMD installations, a permeate flux at a level of a few L/m^{2}h should be expected. For this performance, in order to prevent vapour condensation in the module the air stream in the range of 10–20 m^3/h per 1 m^2 of membranes will be required. Providing such a large gas flow in classic modules with a shell would significantly increase the flow resistance, which can be limited using open capillary modules with lower membrane packing density.

An important point that should be noted is that the application of the capillary modules without an external shell allows to realise this process without feed pre-heating (natural evaporation) with the permeate flux at a level of 0.15–0.3 L/m^{2}h. In the case when hot air (313 K) heating the membrane surface the water evaporation increased to 0.95 L/m^{2}h for feed temperature equal to 302.6 K.

An increase of the feed temperature from 295 K to 330 K allowed to increase the evaporation performance from 0.3 to 2.4 L/m^{2}h for very low gas flow. The evaporation efficiency increased about 60% when gas flow increased to 1.8–2.5 m/s.

Funding: This research was funded by National Science Centre, Poland, grant number 2018/29/B/ST8/00942.

Institutional Review Board Statement: Not applicable.

Informed Consent Statement: Not applicable.

Data Availability Statement: The data presented in this study are available upon request from the corresponding author. The data are not publicly available due to the institutional repository being under construction.

Conflicts of Interest: The author declares no conflict of interest. The funders had no role in the design of the study; in the collection, analyses, or interpretation of data; in the writing of the manuscript, or in the decision to publish the results.

References

1. Li, G.; Lu, L. Modeling and performance analysis of a fully solar-powered stand-alone sweeping gas membrane distillation desalination system for island and coastal households. *Energy Convers. Manag.* **2019**, *205*, 112375. [CrossRef]
2. Janajreh, I.; Suwwan, D.; Hashaikeh, R. Assessment of direct contact membrane distillation under different configurations, velocities and membrane properties. *Appl. Energy* **2017**, *185*, 2058–2073. [CrossRef]
3. Siyal, M.I.; Lee, C.-K.; Park, C.; Khan, A.A.; Kim, J.-O. A review of membrane development in membrane distillation for emulsified industrial or shale gas wastewater treatments with feed containing hybrid impurities. *J. Environ. Manag.* **2019**, *243*, 45–66. [CrossRef] [PubMed]
4. Evans, L.; Miller, J. *Sweeping Gas Membrane Desalination Using Commercial Hydrophobic Hollow Fiber Membranes*; Sandia National Laboratories: Albuquerque, NM, USA, 2002.
5. Boukhriss, M.; Ben Hmida, M.B.; Maatoug, M.A.; Zarzoum, K.; Marzouki, R.; Ben Bacha, H. The design of a unit sweeping gas membrane distillation: Experimental study on a membrane and operating parameters. *Appl. Water Sci.* **2020**, *10*, 1–14. [CrossRef]
6. Karanikola, V.; Corral, A.F.; Jiang, H.; Sáez, A.E.; Ela, W.P.; Arnold, R.G. Sweeping gas membrane distillation: Numerical simulation of mass and heat transfer in a hollow fiber membrane module. *J. Membr. Sci.* **2015**, *483*, 15–24. [CrossRef]

7. Gryta, M. The Application of Submerged Modules for Membrane Distillation. *Membranes* **2020**, *10*, 25. [CrossRef] [PubMed]

8. Duong, H.; Cooper, P.; Nelemans, B.; Cath, T.Y.; Nghiem, L.D. Evaluating energy consumption of air gap membrane distillation for seawater desalination at pilot scale level. *Sep. Purif. Technol.* **2016**, *166*, 55–62. [CrossRef]

9. Rivier, C.; García-Payo, M.; Marison, I.; von Stockar, U. Separation of binary mixtures by thermostatic sweeping gas membrane distillation I. Theory and simulations. *J. Membr. Sci.* **2002**, *201*, 1–16. [CrossRef]

10. Li, G.-P.; Zhang, L.-Z. Laminar flow and conjugate heat and mass transfer in a hollow fiber membrane bundle used for seawater desalination. *Int. J. Heat Mass Transf.* **2017**, *111*, 123–137. [CrossRef]

11. Shokri-Kuehni, S.M.S.; Rad, M.N.; Webb, C.; Shokri, N. Impact of type of salt and ambient conditions on saline water evaporation from porous media. *Adv. Water Resour.* **2017**, *105*, 154–161. [CrossRef]

12. Edwie, F.; Chung, T.-S. Development of simultaneous membrane distillation–crystallization (SMDC) technology for treatment of saturated brine. *Chem. Eng. Sci.* **2013**, *98*, 160–172. [CrossRef]

13. Elsheniti, M.B.; Elbessomy, M.O.; Wagdy, K.; Elsamni, O.A.; Elewa, M.M. Augmenting the distillate water flux of sweeping gas membrane distillation using turbulators: A numerical investigation. *Case Stud. Therm. Eng.* **2021**, *26*, 101180. [CrossRef]

14. Shirazi, M.M.A.; Kargari, A.; Bastani, D.; Soleimani, M.; Fatehi, L. Study on Commercial Membranes and Sweeping Gas Membrane Distillation for Concentrating of Glucose Syrup. *J. MSR* **2020**, *6*, 47–57. [CrossRef]

15. Perfilov, V.; Fila, V.; Marcano, J.S. A general predictive model for sweeping gas membrane distillation. *Desalination* **2018**, *443*, 285–306. [CrossRef]

16. Mousavi, S.A.; Aboosadi, Z.A.; Mansourizadeh, A.; Honarvar, B. Surface modified porous polyetherimide hollow fiber membrane for sweeping gas membrane distillation of dyeing wastewater. *Colloids Surf. A Physicochem. Eng. Asp.* **2020**, *610*, 125439. [CrossRef]

17. Gao, L.; Zhang, J.; Gray, S.; Li, J.-D. Experimental study of hollow fiber permeate gap membrane distillation and its performance comparison with DCMD and SGMD. *Sep. Purif. Technol.* **2017**, *188*, 11–23. [CrossRef]

18. Bagger-Jørgensen, R.; Meyer, A.S.; Pinelo, M.; Varming, C.; Jonsson, G. Recovery of volatile fruit juice aroma compounds by membrane technology: Sweeping gas versus vacuum membrane distillation. *Innov. Food Sci. Emerg. Technol.* **2011**, *12*, 388–397. [CrossRef]

19. Criscuoli, A. Thermal Performance of Integrated Direct Contact and Vacuum Membrane Distillation Units. *Energies* **2021**, *14*, 7405. [CrossRef]

20. Safi, N.N.; Ibrahim, S.S.; Zouli, N.; Majdi, H.S.; Alsalhy, Q.F.; Drioli, E.; Figoli, A. A Systematic Framework for Optimizing a Sweeping Gas Membrane Distillation (SGMD). *Membranes* **2020**, *10*, 254. [CrossRef]

21. Khayet, M.; Cojocaru, C.; Baroudi, A. Modeling and optimization of sweeping gas membrane distillation. *Desalination* **2012**, *287*, 159–166. [CrossRef]

22. Zhao, S.; Feron, P.H.; Xie, Z.; Zhang, J.; Hoang, M. Condensation studies in membrane evaporation and sweeping gas membrane distillation. *J. Membr. Sci.* **2014**, *462*, 9–16. [CrossRef]

23. Abejón, R.; Saidani, H.; Deratani, A.; Richard, C.; Sánchez-Marcano, J. Concentration of 1,3-dimethyl-2-imidazolidinone in Aqueous Solutions by Sweeping Gas Membrane Distillation: From Bench to Industrial Scale. *Membranes* **2019**, *9*, 158. [CrossRef]

24. Huang, S.-M.; Chen, Y.-H.; Yuan, W.-Z.; Zhao, S.; Hong, Y.; Ye, W.-B.; Yang, M. Heat and mass transfer in a hollow fiber membrane contactor for sweeping gas membrane distillation. *Sep. Purif. Technol.* **2019**, *220*, 334–344. [CrossRef]

25. Gryta, M.; Tomaszewska, M.; Morawski, A.W. A Capillary Module for Membrane Distillation Process. *Chem. Pap.* **2000**, *54*, 370–374.

26. Zhao, S.; Wardhaugh, L.; Zhang, J.; Feron, P. Condensation, re-evaporation and associated heat transfer in membrane evaporation and sweeping gas membrane distillation. *J. Membr. Sci.* **2015**, *475*, 445–454 [CrossRef]

27. Safi, M.A.; Prasianakis, N.; Mantzaras, J.; Lamibrac, A.; Büchi, F.N. Experimental and pore-level numerical investigation of water evaporation in gas diffusion layers of polymer electrolyte fuel cells. *Int. J. Heat Mass Transf.* **2017**, *115*, 238–249. [CrossRef]

28. Gryta, M. Fouling in direct contact membrane distillation process. *J. Membr. Sci.* **2008**, *325*, 383–394. [CrossRef]

29. Shirazi, M.M.A.; Kargari, A.; Tabatabaei, M.; Ismail, A.F.; Matsuura, T. Concentration of glycerol from dilute glycerol wastewater using sweeping gas membrane distillation. *Chem. Eng. Process. Process. Intensif.* **2014**, *78*, 58–66. [CrossRef]

30. Nakoa, K.; Rahaoui, K.; Date, A.; Akbarzadeh, A. Sustainable zero liquid discharge desalination (SZLDD). *Sol. Energy* **2016**, *135*, 337–347. [CrossRef]

31. Guan, Y.; Li, J.; Cheng, F.; Zhao, J.; Wang, X. Influence of salt concentration on DCMD performance for treatment of highly concentrated NaCl, KCl, MgCl2 and MgSO4 solutions. *Desalination* **2015**, *355*, 110–117. [CrossRef]

32. Shirazi, M.M.A.; Kargari, A. Concentrating of Sugar Syrup in Bioethanol Production Using Sweeping Gas Membrane Distillation. *Membranes* **2019**, *9*, 59. [CrossRef] [PubMed]

33. Duong, H.; Cooper, P.; Nelemans, B.; Cath, T.Y.; Nghiem, L.D. Optimising thermal efficiency of direct contact membrane distillation by brine recycling for small-scale seawater desalination. *Desalination* **2015**, *374*, 1–9. [CrossRef]

34. Venture Industries. Available online: https://Ventur.Eu (accessed on 1 February 2022).

Article

Performance Study of Eductor with Finite Secondary Source for Membrane Distillation

Ravi Koirala [†], Xing Zhang [†], Eliza Rupakheti [†], Kiao Inthavong [†] and Abhijit Date *,[†]

School of Engineering, RMIT University, Melbourne, VIC 3083, Australia
* Correspondence: abhijit.date@rmit.edu.au
† These authors contributed equally to this work.

Abstract: This is an experimental work performed to identify the influence of direct contact condensation inside an eductor. The fluid used in the experiments is water in two different phases: liquid and vapor, for primary and secondary flows, respectively. This study was conducted in an attempt to establish the suitability of an eductor as a combined vacuum generator and condenser for membrane desalination applications. The pressure and temperature measurements at critical points in the flow paths have been summarized to identify the influence of primary flow on secondary fluid saturation and condensation. In addition, the mechanism of phase change has been explained through the photography of fluid flow in a two-dimensional eductor. A consistent oscillation of the gas-liquid interface was observed during steady-state operations of the eductor. This work also contributes to the validation of future computational research. It will provide a baseline for computational thermal fluid analysis related to the mixing of condensing and non-condensing flow. In general, the research encompasses the practical operational scenario and provides information on the heat and mass transfer of direct contact condensation with a finite secondary source.

Keywords: eductor; condensation; saturation; experiment; mechanism visualization

Citation: Koirala, R.; Zhang, X.; Rupakheti, E.; Inthavong, K.; Date, A. Performance Study of Eductor with Finite Secondary Source for Membrane Distillation. *Energies* **2022**, *15*, 8620. https://doi.org/10.3390/en15228620

Academic Editor: Alessandra Criscuoli

Received: 23 October 2022
Accepted: 14 November 2022
Published: 17 November 2022

Publisher's Note: MDPI stays neutral with regard to jurisdictional claims in published maps and institutional affiliations.

1. Introduction

Ejectors/eductors are mechanical devices that operate on the principle of converting primary flow energy into secondary fluid entrainment. They are among the standard industrial devices used for moving liquid, particles, and gas or to create a vacuum. The principal fluid might be either liquid or gas, depending on the application [1,2]. They can also perform multiple fluid dynamic features like multiphase mixing, heat and mass transfer, pumping, fluid flow expansion, and compression [3–5]. The absence of moving parts and simple construction make them some of the more widely preferred multifunctional components [6].

The existing literature outlines the role of the eductor and how it might be modified for different uses. Most of the existing literature focuses on air jet or steam jet systems, with only a few studies conducted on water-jet eductors. Zhang et al. [7] performed an experimental and empirical study of flow inside a water jet eductor. The secondary source used in this study was superheated steam from a steam generator. Axial pressure measurements were performed to explain the phenomenon. The pressure along the axial direction decreased up to the exit of the throat for the selected geometry. Shah et al. [8] conducted a numerical and experimental study of a steam jet ejector with the primary fluid being superheated steam and the secondary fluid being cooling water. Their numerical study considered an infinite source of liquid, which is not difficult to approximate, since water is in liquid form under standard conditions. This is among the limited studies considering the mass transfer rate during the computational analysis of ejectors. Banu et al. [9] performed a numerical and experimental study for a refrigeration application with a primary fluid of superheated steam. The primary focus of the study is to evaluate the influence of primary fluid swirl on the performance of ejectors. An increase in primary fluid swirl was found to have enhanced entrainment. Yan et al. [10] performed an experimental study with swirling primary fluid (water) to entrain steam from the secondary source. These swirling vanes introduced vortices resulting in more significant interaction time between the fluids for better mixing [11,12]. Narabayashi et al. [13] experimentally studied an ejector for a passive core injection system to be used in next-generation reactors, with primary water, and secondary steam fitted with throat drains. The use of steam from a high-pressure turbine to run the steam injector reduced the plant efficiency by 1%. Yan et al. [14] performed experiments with primary steam and secondary water for heating purposes. A shock wave was reported to occur during phase change within the flow channel. One of the key factors in the existing literature is the use of an infinite and common secondary source (e.g., superheated steam, compressed air, or water). Some of the practical aspects that have been identified as significant knowledge gaps are the behavior of eductor-aided sub-atmospheric saturation conditions, low-grade heat input for phase transformation, and synchronous multifunctional activity in eductors (e.g., combined degassing, pumping, and heat exchange). In addition, for applications that include 2 phase single species flows, eductors can be developed to function as a direct contact heat transfer device/condenser.

This work focuses on two objectives: First, determining the influence of eductor operating conditions on the sub-atmospheric vapor generation at the secondary flow source and condensation within the eductor. Second, establishing a foundation for the verification of computational studies on the thermo-fluid analysis of similar phenomena. The plan is to develop an eductor for use as an active vapor transfer and condensation unit for desalination systems. During this study, the pressure at different axial positions and the influence of phase change during direct contact condensation were examined, along with the eductor-aided saturation process in a control volume. The thermodynamic mechanism of heat transfer (explored using a T-S diagram), calculation of the overall heat transfer coefficient, and overall exergy analysis are three major theoretical knowledge contributions that can provide a new direction to the existing work. The outcome of this work can also be referenced for verification of computational activities related to the following four thermal-fluid processes: flow dynamics in eductors, heat and mass transfer, multiphase fluid mixing, and sub-atmospheric saturation.

2. Eductors for Active Vapor Transport and Condensation

The primary and secondary inlet, as well as the suction chamber, mixing chamber, throat, and diffuser, make up an eductor. Eductors work by converting pressure energy into velocity energy, with secondary entrainment occurring in the low-pressure zone. When a high-velocity primary fluid leaves the primary nozzle, it generates a continuous eddy current in the surrounding fluid, keeping the suction chamber at low pressure. When connected to a secondary fluid source, the secondary port degasses the area between the secondary fluid surface and the mixing chamber before pulling it to the mainstream. The physical model of an eductor for active vapor transfer and condensation is shown in Figure 1. The input work supply pump, condenser, and pressure rising diffuser are the main functional components. The working fluid, in this case, is liquid water, which serves numerous roles during the energy conversion. Liquid works as a working fluid in the pumping portion, pulling secondary fluid (water vapor) into the mainstream. This flow is caused by a positive pressure difference between the secondary fluid source and the suction chamber. In addition, it works as a condensing/cooling fluid in the condenser by direct contact condensation and cooling. A single-phase fluid made up of the primary and secondary flows exits the eductor system through the diffuser at a pressure higher than the secondary flow inlet but lower than the primary flow inlet. The greater the rate of condensation, the greater the entrainment because the collapsing vapor bubbles resulting from condensation form large empty volumes, allowing for additional incoming flow to fill the space. Outlet pressure is a limiting component; raising the outlet pressure improves the condensation rate but obstructs the fluid flow; thus, entrainment rises for a brief period before progressively decreasing.

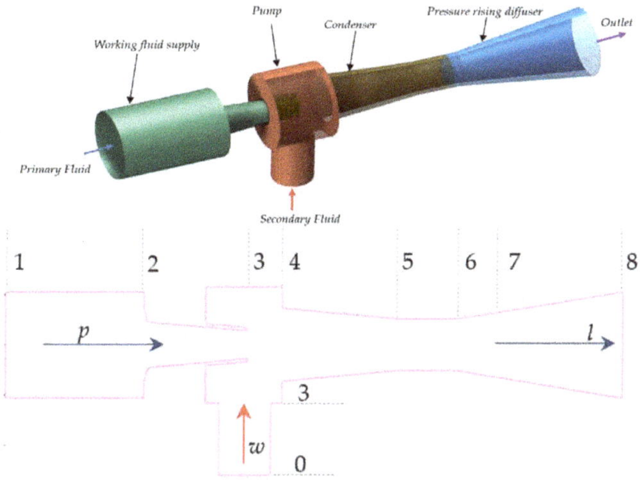

Figure 1. Physical model of eductor.

The performance of the eductor is defined in terms of its entrainment ratio and pressure ratio. The entrainment ratio is defined as the ratio of the mass flow rate of secondary fluid that is pumped into the system to the mass flow rate of the primary fluid that is driving the pumping action, given by Equation (1).

$$Entrainment\ ratio\ (E_r) = \dot{m}_s / \dot{m}_i \tag{1}$$

The pressure ratio is defined as the mechanical compression capacity of the entrained vapor, given by Equation (2).

$$Pressure\ ratio\ (P_r) = (P_o - P_s)/(P_i - P_s) \tag{2}$$

Figure 2 is the estimated T-s diagram for a two-phase single species flow, with condensing secondary flow. The high velocity subcooled liquid (primary fluid) reaches the suction chamber at a constant temperature and slightly higher entropy due to its low pressure. As a result of the pressure difference between the secondary source and the suction chamber, saturated vapor (secondary fluid) enters the system from 3w. Inside the chamber, the two phases interact with each other and the energy exchange is initiated. At the secondary fluid inlet, the energy balance between the vapor source and the suction chamber can explain the vapor rising phenomenon in the eductor. Applying this energy balance, the secondary mass flow rate can be described through Equation (3):

$$\dot{m}_s = \rho_v\, A_s\, \sqrt{(2(h_{vo} - h_{v1}))} \tag{3}$$

In addition, the primary fluid mass flow rate can be estimated by continuity and the Bernoulli equation applied to the primary nozzle:

$$\dot{m}_i = \rho_l A_n \sqrt{2\,\eta_n \left(\frac{P_{l0} - P_2}{\rho_l} \right)} \tag{4}$$

The velocity of the subcooled liquid is initially much higher than the saturated vapor, hence the effect of the friction between the phases at their interface propels the movement of the secondary fluid. This shear stress between the phases due to leading and lagging velocity is given by:

$$\tau = \frac{1}{2} f \rho_v (V_v - V_l)^2 \tag{5}$$

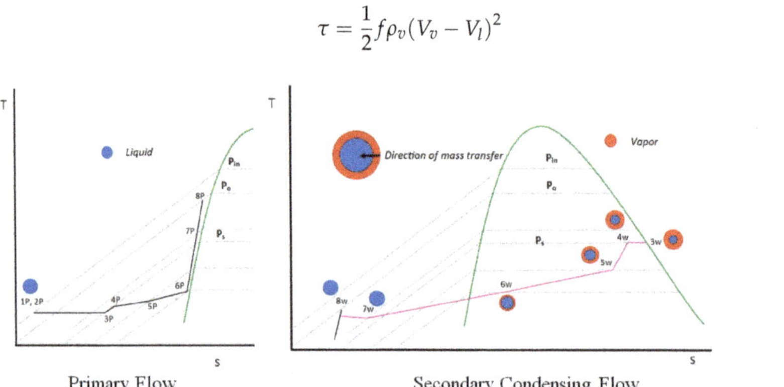

Primary Flow   Secondary Condensing Flow

Figure 2. T-S diagram estimated for the eductor.

With the thermal interaction between the two phases, the vapor fraction starts decreasing in the annular flow surrounding the primary fluid flow. The sub-cooled liquid continues to receive thermal energy (the latent heat of vaporization) released during the phase change. The pressure energy in the fluid also varies due to the variation in the axial cross section of the eductor. By the time, the primary and secondary fluids reach the throat, complete condensation takes place, and the two streams finally form a homogeneous mixture in the diffuser (between points 7 and 8). Although the local heat transfer coefficient at a given axial location, in this case, is a function of the local pressure and the inter-phase area between Sections 4–6, the overall coefficient of heat transfer can be defined by:

$$H_c = \frac{\dot{m}_v \, h_{fg}}{(T_v - T_l) \, A_{iA}} \tag{6}$$

where H_c is the overall convective heat transfer coefficient, $\dot{m}_v$ is vapor mass flow rate, $_{fg}$ is the latent heat of condensation, T_v is vapor temperature (secondary fluid), T_l is the liquid temperature (primary fluid), and A_{iA} is the total surface area of interphase interface.

3. Experimental Description

Figure 3 is the schematic view of the components for the experimental setup, together with the position of the sensors and the instrumentation. The major components of the setup include a centrifugal pump, pressure tank, eductor, viewing piece, sump, piping, and instrumentation for data acquisition and recording. A piezo-resistive sensor for suction pressure measurement and pressure transducers for the pressures at other locations are connected to a Datataker DT80, along with thermocouples for temperature measurements. In open loop operation, the primary flow sub-cooled liquid water is pumped using a centrifugal pump to the inlet of the eductor and the outlet is discharged to the sump. The setup can also function in closed-loop operation for particle tracking and flow visualization. The low-pressure secondary vapor is generated by supplying heat to a vacuum-sealed conical flask via a plate heater. The power rating of the plate heater is described in Table 1. The reliability of the experimental setup has been ensured through sensor calibration, repeatability testing, and uncertainty analysis. The relative uncertainty of the sensors used in the system is in the range of ±0.4% to 1.2%.

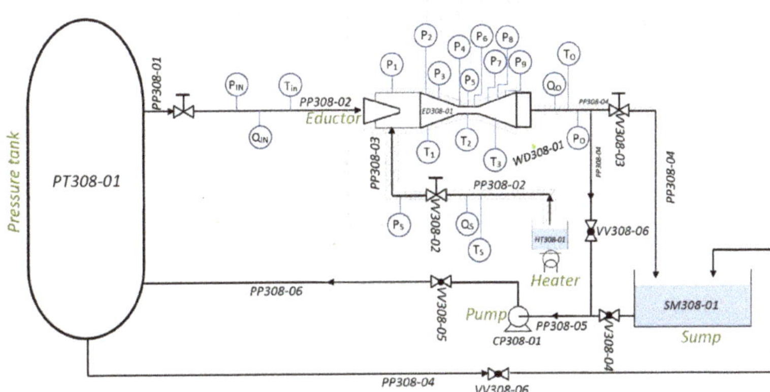

Figure 3. Schematic diagram of experimental setup.

Table 1. Operational parameters of Eductor.

Parameter	Description						Units
Inlet Pressure	168.3						kPa
Inlet Temperature	22.4						°C
Inlet Flow rate	16.54						LPM
Outlet Pressure	BP$_1$	BP$_2$	BP$_3$	BP$_4$	BP$_5$	BP$_6$	kPa
	98.67	100.12	101.99	106.31	111.70	118.83	
Suction Temperature	BP$_1$	BP$_2$	BP$_3$	BP$_4$	BP$_5$	BP$_6$	°C
	93.61	94.54	95.26	96.48	98.37	105.1	

Figure 4 is the experimental setup built at the Thermo-Fluid lab, RMIT University, Australia for performance estimation of eductors. For this experiment, the primary flow creates a low pressure in the space between the water surface in the conical flask (Figure 4d) and the eductor suction chamber, hence allowing the secondary fluid in the flask to reach saturation at lower temperatures. The generated vapor is entrained into the mainstream of the eductor where the exchange of heat and mass occurs between the primary and secondary fluid streams. To maintain consistency, all the experiments were conducted at a constant heat supply rate from the plate heater and a constant inlet pressure. The valve VV308-02 was fully opened and VV308-06 was fully closed. Eductors operate on the pressure difference between their inlet and outlet; therefore, given the constant inlet pressure condition, the experiment was performed by controlling the outlet pressure, which in turn varied the suction pressure of the eductor. Six different opening levels of VV308-03 were selected to measure the influence of backpressure on performance. The operational conditions have been described in Table 1. The axial pressure was measured using pressure taps in three different axial positions (suction chamber, mixing chamber, and throat) (Figure 4a). The eductor used for this study is a 3D printed part reverse-engineered from a commercial eductor design and modified to accommodate three pressure ports for experiments (see Figure 4a).

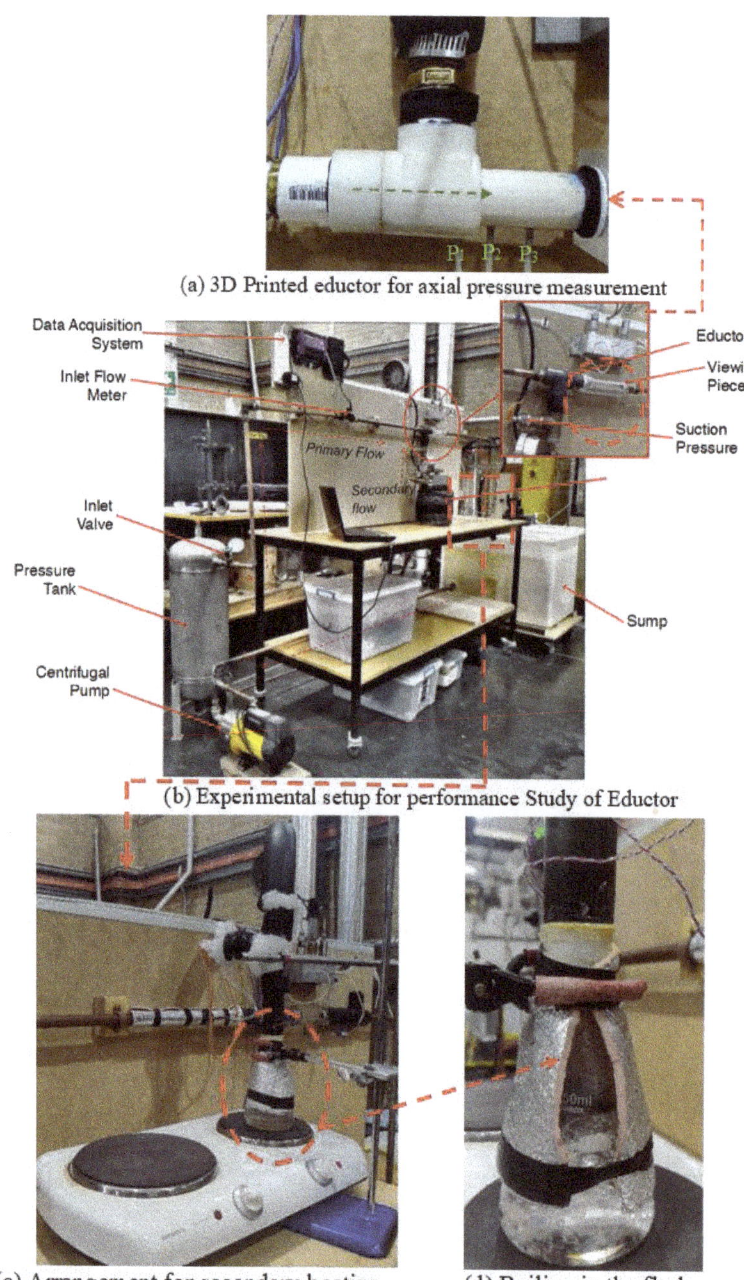

Figure 4. Experimental setup at Thermo-fluid Lab-RMIT.

4. Results and Discussion

This section describes the experimental results with regard to three different criteria: thermal saturation phenomenon, axial flow in non-condensing and condensing cases, and functional performance of the eductor (i.e., working range for outlet pressure and entrainment ratio).

4.1. Sub-Atmospheric Vapor Generation

In the experimental setup (Figures 4 and 5), the secondary flow of water vapor was produced by heating liquid water inside a sealed conical flask using a plate heater, while the pressure inside the flask was controlled by the eductors suction action to maintain sub-atmospheric conditions. This section describes the thermal saturation phenomenon and the influence of the eductors parameters on the process. This is one of the contrasting features of this study, i.e., where the response of an eductor to a finite secondary source has been evaluated. In practice, an unlimited vapor source will be of no practical interest during hybrid desalination. Figure 5 is a schematic of the conversion of mechanical and thermal energy within an eductor. The mechanical energy in the primary fluid is used to maintain low pressure and pumping of the secondary fluid. Similarly, the thermal energy contributes to saturated vapor generation and direct contact condensation (DCC) of the secondary fluid.

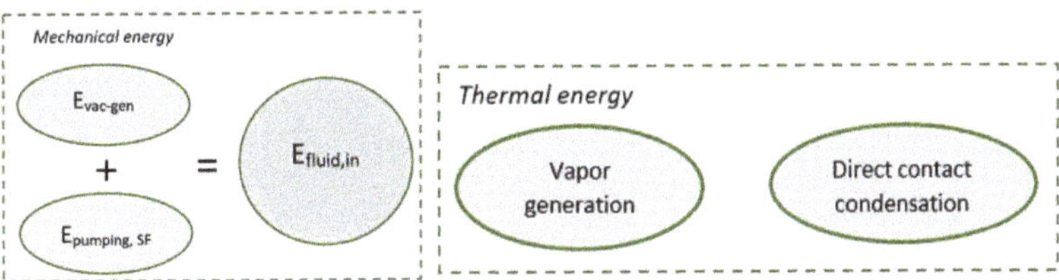

Figure 5. Fluid energy scenario in the eductor.

There are two thermocouples fitted in the vapor generation apparatus of the experimental setup to monitor the conditions during evaporation of the secondary fluid: the first one submerged in the liquid water towards the base of the flask and the second one suspended near the mouth of the flask. The measurement of suction pressure was performed at about 29 cm above the mouth of the flask. In the following section, the key observations made during the experiments regarding the secondary fluid evaporation under sub-atmospheric conditions with respect to the operational characteristic of the eductor have been discussed.

Figures 5 and 6 show secondary flow (suction) pressure, secondary fluid liquid temperature, and secondary fluid vapor temperature measurements successively as the secondary fluid changes from subcooled to saturated condition, while the primary flow is maintained at back pressure 1 (BP_1). In Figures 5 and 6, before the start of active vapor transfer (i.e., during secondary fluid heating), the pressure fluctuations were monitored in the suction chamber, with an average pressure of 83 kPa being recorded. Although the secondary fluid temperature was well below the saturation temperature corresponding to the measured suction pressure, there was no vapor generation and, therefore, no vapor transfer. The vapor generation only started when the secondary fluid temperature reached the saturation condition, allowing active vapor transfer to occur between the vapor generation source and the eductor. It can be seen in Figures 5 and 6 that there was a period of abrupt pressure drop, known as coughing flow, which represents the period in which a small amount of vapor is generated, pushing the small amount of air present in the flask to the suction chamber, which in turn is extracted by the eductor. This removal of air, while the secondary fluid has not reached saturation temperature, causes the suction pressure to drop for a short period, as can be seen in the figure. This coughing effect is observed for almost all operating conditions and is one of the important start-up considerations that large-scale applications need to consider to prevent collapsing of the secondary fluid pipe due to the sudden vacuum created.

After the secondary fluid reaches thermal saturation, the vapor flow starts, and the average suction pressure stabilizes at 85 kPa. Compared to when the secondary fluid is sub-cooled, large fluctuations in suction pressure are observed when the secondary fluid reaches thermal saturation. The accompanying small fluctuations in temperature could be due to these changes in pressure causing a change in the saturation point. This fluctuation in the suction pressure is due to the direct condensation of water vapor in the sub-cooled primary flow (liquid water). The vapor bubbles from the secondary fluid flow enter the sub-cooled primary flow and abruptly collapse due to rapid heat and mass transfer. This gives rise to pressure waves (shock waves) which are recorded as large fluctuations in the suction pressure measurement.

With increasing back pressure in the system, the fluctuations in the suction pressure become even greater. This is due to the shifting of the condensation region towards the mixing chamber, which is closer to the suction port. Similarly, a similar behavior was reported during the computational study by Koirala et al. [15]. The mechanical performance of the system is negatively affected by the back pressure; however, an improvement in thermal performance has been observed. The study shows that for two phase single-species condensing flow, thermal activity has a greater contribution to entrainment compared to mechanical work. The increase in condensation rate with larger interaction time allows more vapor to enter the system, but this is limited to the critical point as a certain amount of mechanical work is essential to accelerate the process. The effect on pressure and entrainment has been elaborated on in the following sections.

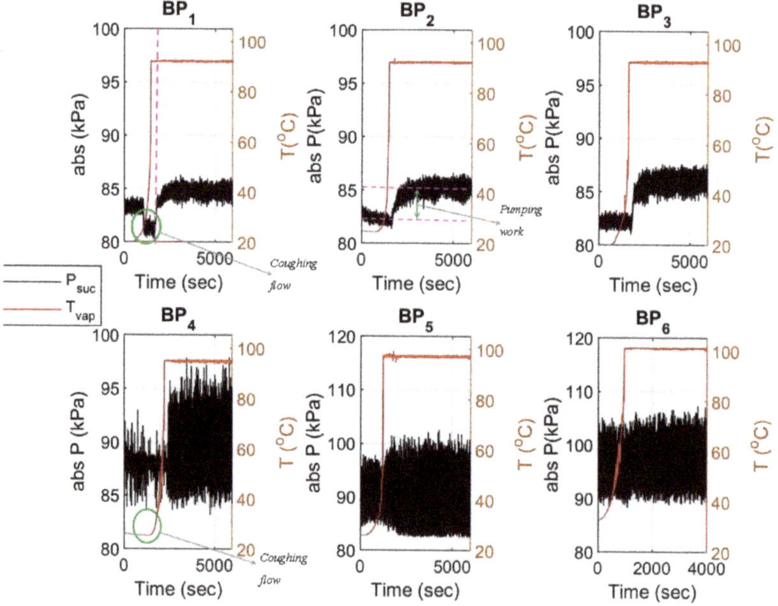

Figure 6. Pressure and temperature measured at secondary source for different back pressures.

4.2. Axial Pressure Distribution

4.2.1. Non-Condensing Flow

Figure 7 is the pressure distribution measurement over the axial points P_1, P_2, and P_3 (as shown in Figure 4a at different back pressures for secondary non-condensing flow (air)). The lowest pressures for P_1, P_2, and P_3 were measured at BP_2 conditions with values of approximately 70 kPa, 75 kPa, and 90 kPa (absolute) respectively. Similarly, the highest pressures for P_1, P_2, and P_3 were measured at BP_6 condition with values of 92 kPa, 98 kPa, and 110 kPa (absolute), respectively. In an eductor, the two-phase region with the lowest pressure is at point P_1, which is measured via a tapping fitted in the wall of the suction chamber. This is the region immediately after the primary nozzle where the two phases (primary and secondary fluids) first come in contact and start to mix. The design and arrangement of this component play an important role in the entrainment capacity of an eductor. The most suitable positioning will maintain the largest possible pressure difference between the secondary source and the eductor flow path, resulting in the highest level of entrainment. With increasing backpressure, the pressure at each of these points was measured and found to increase.

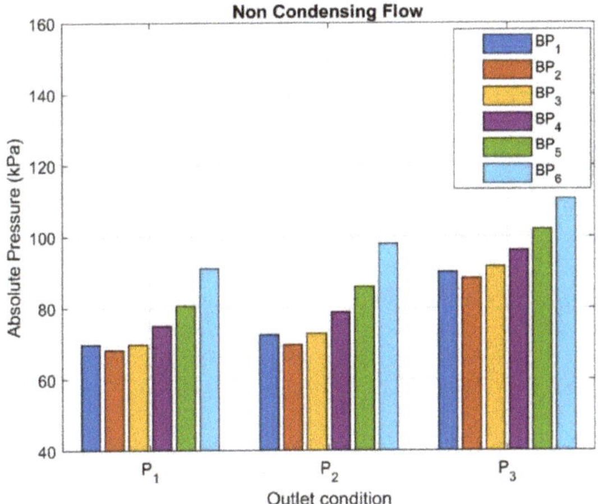

Figure 7. Axial pressure distribution in non-condensing flow.

4.2.2. Condensing Flow

Figure 8 is the graph for the time-averaged pressure distribution measurements at the axial points P_1, P_2, and P_3 for different back pressures, with secondary condensing flow (water vapor), at the same conditions defined in the previous section. For each of the points, the lowest pressure was measured at the back pressure BP_1, and highest pressure was measured for the back pressure BP_6. The lowest pressures of 69 kPa, 72 kPa, and 87 kPa (absolute) were measured under conditions P_1, P_2, and P_3 respectively at BP_1 condition. Similarly, the highest-pressure values of 92 kPa, 98 kPa, and 108 kPa (absolute) were measured at P_1, P_2, and P_3, respectively, at BP_6 condition. The impact of back pressure was found to be greater at P_2 compared to P_1 and P_3.

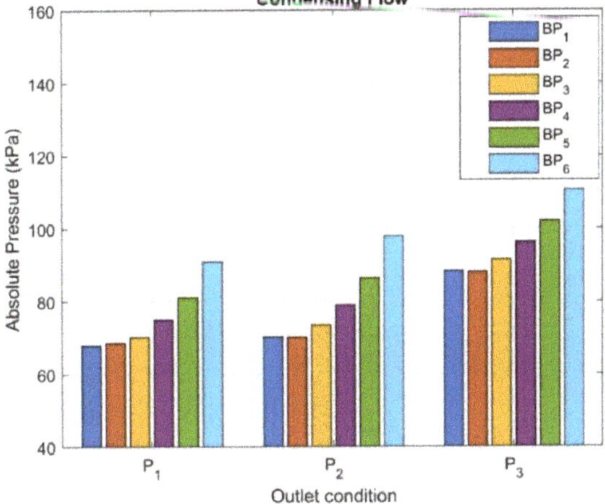

Figure 8. Axial pressure distribution in condensing flow.

4.3. Influence of Condensation

4.3.1. Quantitative Study

Figure 9 shows the comparative study of pressure fluctuation between condensing and non-condensing cases for different back pressure conditions at points P_1, P_2, and P_3. In most of the experiments related to phase change, condensation was reported to result in large bubbles collapsing. They are either visually monitored or measured through pressure fluctuations. Here, a statistical method based on the standard deviation was selected to summarize the variation in the crest and trough from mean pressure values. The cases with the non-condensing flow can be considered as controls or references, where, except for some minor condensation of atmospheric moisture, the air is largely dry. The condensing flow contains a finite volume of water vapor. In the case of BP_1, i.e., the lowest back pressure, the largest fluctuation was measured at P_3 (in the throat); hence, most of the mass transfer occurred there. For the cases of BP_2, BP_3, and BP_4, the fluctuation tends to stabilize due to the damping action of the increased back pressure. In addition, the largest fluctuation was also found to shift toward P_2, indicating that most of the condensation starts to occur in this region. For the cases of BP_5 and BP_6, with further increased damping, most of the condensation was measured to occur in the suction chamber and the mixing chamber. The influence of this is also reported in Figure 6.

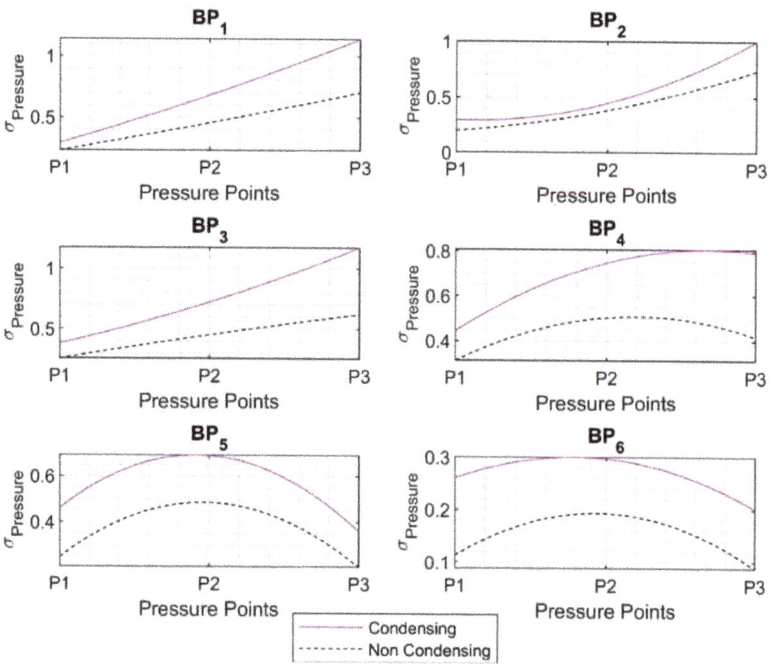

Figure 9. Deviation from the mean pressure.

4.3.2. Qualitative Study of Eductor Mechanism

Figure 10 shows the adjustments made to the experimental setup to visually observe if complete condensation occurs within the system. On the basis of the prior literature, it would be expected that there would be bubbles in the water stream in case of unfinished condensation. During the entire series of experiments; however, no bubbles were observed beyond this point. Therefore, it can be assumed that under the operational conditions applied, complete condensation occurs within the eductor system. It is necessary, however, to find the mechanism of mass transfer over the axial region of the inter-phase interaction within the eductor.

Figure 10. Visual inspection of bubbles.

The answer to this question regarding the mechanism of mass transfer helps to strengthen the understanding related to internal flow. There is evidence of visualization practices for internal flows in the existing literature. Yang et al. [16] studied the condensation of steam in a sub-cooled water pool using a nozzle and barrel adjustment. Innings et al. [17] visually studied steam condensation in an ultra-high temperature treatment system.

The current study developed a cascade geometry, Figure 11, to make visual records of the condensing flow of low-pressure vapor into a sub-cooled water jet. The Figure 12 is the pressure distribution plotted based on the computational analysis. The setup was designed such that the pressure distribution within the cascade channel becomes like the eductor. It is analogous to the axisymmetric geometry of an eductor, extruded to have a similar wall pressure. The flow video was taken at a frame rate of 480 FPS. There are records of theoretical descriptions of flow within an eductor, but visual evidence of actual flow is missing. This is the first experimental evidence showing the mechanism of mass transfer between low-pressure secondary fluid vapor and primary fluid water jet. Figure 13 is a summary of the images showing two-phase mixing and mass transfer within the cascade. It includes two phase regions between the suction chamber and the diffuser. At the primary fluid volume flow rate of 6.2 LPM and vapor temperature of 98.24 °C, the low-pressure vapor slowly entrains into the main flow stream. The thick mixing region can be clearly seen, where the saturated vapor is in transition to reach the compressed liquid. The vapor axially oscillates between the mixing chamber and throat. The axial pressure is a function of back pressure, in this case, it gradually increases from the suction chamber to the diffuser [18]. Hence the saturated vapor initially entrained in the passage starts condensing through this non-constant cross-sectional mixing chamber. At the same time, the vapor temperature is also gradually decreasing, and hence the point of complete condensation shifts to the start of the mixing chamber. The varying area of the vapor–liquid interface and the resulting variation in the heat and mass transfer coefficient results in the oscillations observed. This is also visible in the pressure fluctuations seen in Figure 6.

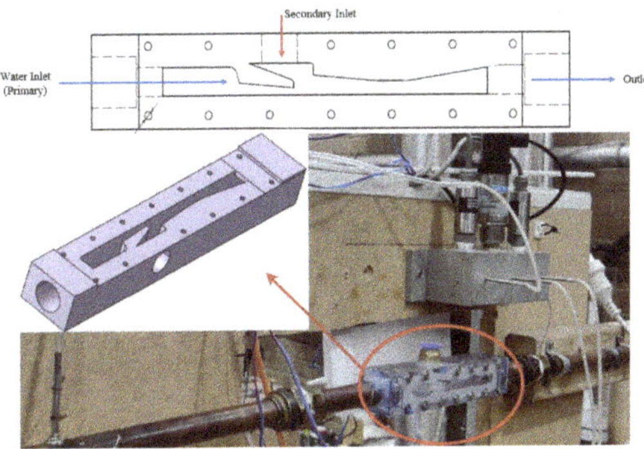

Figure 11. 3D printed transparent cascade.

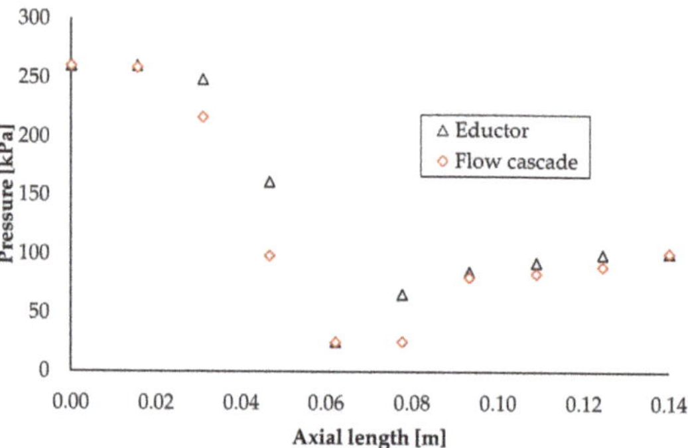

Figure 12. The pressure distribution between eductor and cascade for the analogy of the study.

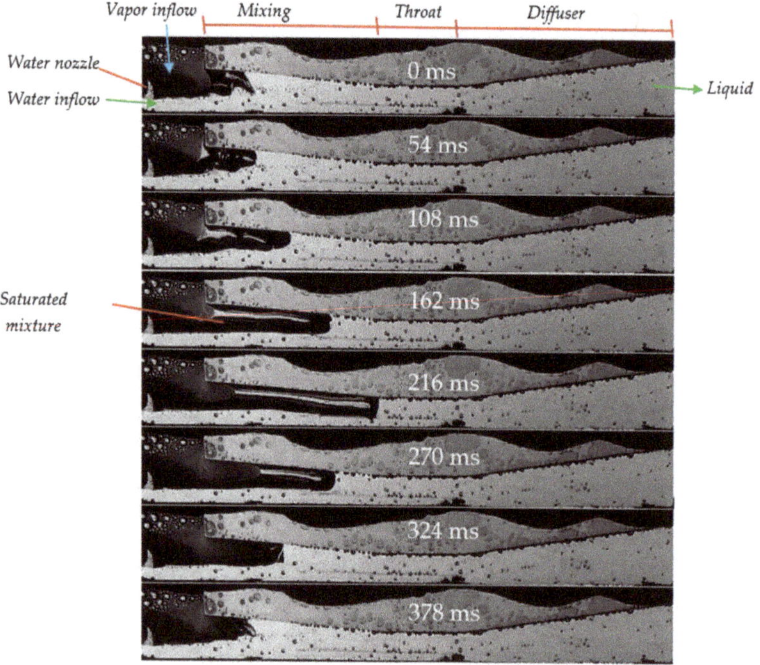

Figure 13. Framewise observation of phase change within Eductor cascade.

4.4. Performance of Eductor

4.4.1. Entrainment Ratio

Figure 14 shows the plot for the entrainment ratio with respect to the pressure ratio of the eductor. Here, the pressure ratio is defined as the ratio of the difference between the outlet and suction pressure to the difference between inlet and suction pressure (see Equation (5)-1). With increasing pressure ratio, the entrainment ratio was measured to decrease. The highest entrainment ratio of 0.0001452 was recorded for a pressure ratio of 0.17. The measurements were taken until the condition in Figure 14 was achieved. During the experiments, the pressure ratio was controlled by controlling the back pressure (outlet pressure of the Po system). The momentum transfer from the high-velocity jet to the surrounding fluid in the suction chamber performs degassing and creates a low-pressure zone. The secondary fluid at higher pressure flows inward to maintain equilibrium. In the case of a control volume with a finite secondary fluid source, the available secondary fluid is pumped in, and the remaining energy is used to maintain the sub-atmospheric pressure. The total energy is divided into entraining the secondary flow and maintaining the system's sub-atmospheric condition. The entrainment ratio and pressure ratio are the dimensionless numbers defining these two functional characteristics of an eductor. From the Equations (1) and (2), this can further aid in visualizing the characteristics of an eductor as both a compressor and a pump during simultaneous operation.

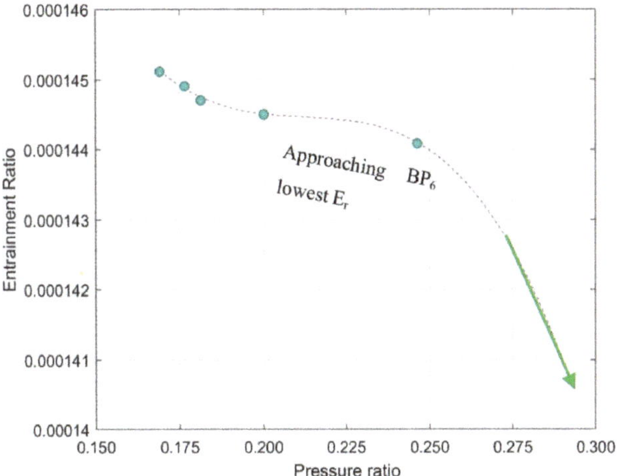

Figure 14. Entrainment ratio vs. Pressure ratio for eductor.

4.4.2. Maximum Operational Range

Figure 15 shows the optimal operating conditions (points) of the eductor at different inlet pressures. It is at the value of back pressure (outlet pressure Po) where there will be no secondary entrainment. Any further increase in the outlet (back) pressure value will result in reverse flow through the secondary nozzle. The respective flow rate has also been plotted to visualize the total energy of the system. With increasing inlet pressure, the difference between inlet pressure and the ultimate back pressure point is larger. The conversion of energy as the primary fluid moves axially forward is the prime operational characteristic. The conversion of pressure energy to kinetic energy, and then to pressure energy with positive action of friction on the interface for the work done is the specific modality of conversion. The value of inlet pressure and corresponding flow rate describe the total inlet energy. The measured value of optimum compressibility (outlet pressure) for the corresponding inlet conditions creates an idea of the operational extent of eductor.

4.4.3. Efficiency and Exergy Analysis

i. Efficiency

As the eductor can perform multiple functions based on its application, e.g., compression, pumping, mixing, condensing, degassing, etc., its efficiencies are also defined accordingly. In this application, defining the eductor as a compressor tends to cover most of its functionality. Therefore, the compression efficiency of the eductor has been defined as the ratio of the rate of work carried out on the secondary fluid to the rate of work supplied by the primary fluid (see Equation (5)-7). The maximum efficiency of 12% was calculated for the eductor operating at BP_5 (Figure 16). This efficiency term incorporates the effect of multiple functions that occur in the background. The work involves degassing the suction chamber, pumping secondary fluid into the chamber and then in the direction of the outlet, exchange of thermal energy and mass between the two fluid streams, and finally releasing at a pressure higher than the suction pressure. The system is analogous to the piston-cylinder system, where the cylinder is the eductor, the piston is the primary fluid, and the volume X is the suction chamber. When the piston moves in the direction A, X becomes a low-pressure zone, as the volume increases but the mass content of the system remains constant. As soon as the valve is opened, the surrounding fluid rushes into volume X. With an eductor, the primary fluid sweeps away the gas content in the suction chamber, converting it to a low-pressure volume analogous to X, as soon as the secondary inlet valve is opened, as the fluid rushes in to fill the space. The pressure difference is the prime driving force within the eductor, which can be further aided by mass transfer due to condensation. The efficiency for other functions (e.g., pumping efficiency, heat, and mass transfer performance, etc.) can be higher or lower, as the definitions of input and output for these functions can vary. The compression efficiency is a good indicator of overall performance, as it is the result of all the mechanisms that occur within an eductor (Equation (7)).

$$\eta_{eductor} = \frac{\dot{V}_{suction} \times (P_{out} - P_{suction})}{\dot{V}_{inlet} \times (P_{in} - P_{out})} \times 100 \tag{7}$$

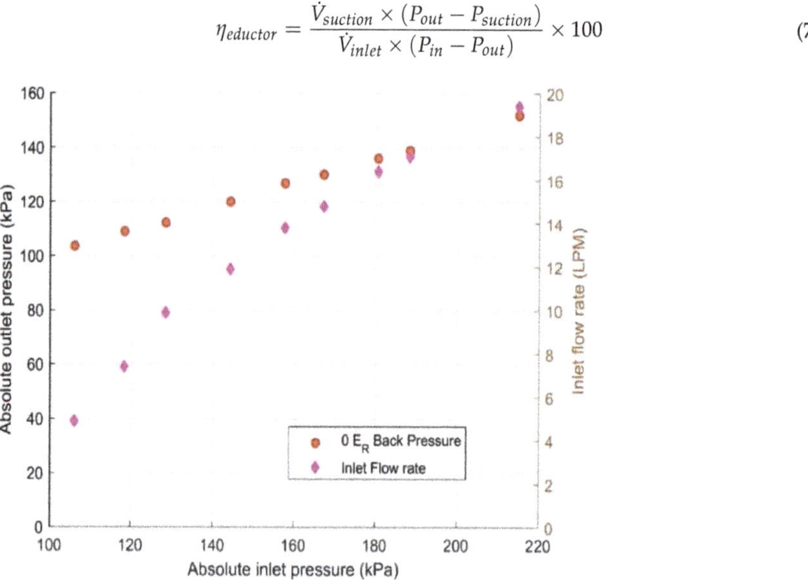

Figure 15. Optimum operational point of eductor.

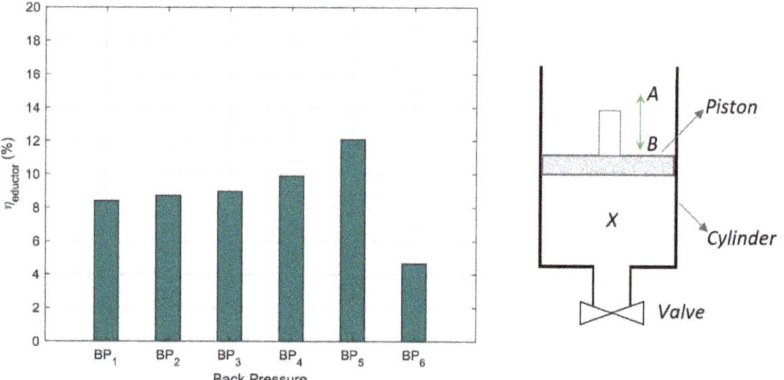

Figure 16. Efficiency of Eductor at different back pressure.

ii. Exergy (xE) analysis

Exergy analysis has been performed to analyze the energy destruction within an eductor system for different back pressures. Equations (8)–(10) define the inlet, suction, and outlet-specific exergy, respectively. The exergy and exergy destruction in the eductor were calculated using Equation (11) and Equation (12), respectively. The dead state pressure and temperatures were taken to be 100 kPa and 295.4 K, respectively. For the exergy analysis, velocities of the primary and secondary fluid have been calculated based on the flow measurements. The velocity of the primary fluid at point 1 was estimated using the volume flow rate, specific volume, and cross-sectional area at point 1 (Figure 17). A vortex flow meter (Grundfos VFS 2-40) was used to measure the volume flow rate of the primary fluid. The velocity of the secondary fluid at point 3 was estimated based on the average mass flow rate, specific volume, and cross-sectional area at point 3. The average mass flow rate of the secondary fluid was estimated based on the change in the mass of the secondary fluid in the boiling flask over the period of the experiment. The specific volume of the secondary fluid was assumed to be equal to a specific volume of saturated water vapor at the suction pressure measured at point 3. The diameter of inlet, outlet, suction, and nozzle are 22 mm, 20 mm, 22 mm, and 5 mm, respectively. Standard water property tables were used to estimate the values of specific enthalpy and specific entropy at different points in the eductors at measured pressures and temperatures [19].

Figure 17. Locations in eductor for exergy (xE) analysis.

The calculation of exergy for difference locations within eductor was calculated using the following equations;

Primary inlet;

$$xe_{1,l} = (h_{1,l} - h_o) - T_o(s_{1,l} - s_o) + \frac{v_{1,l}^2}{2} + gz_{1,l} \tag{8}$$

Secondary Inlet;

$$xe_{3,v} = (h_{3,v} - h_o) - T_o(s_{3,v} - s_o) + \frac{v_{3,v}^2}{2} + gz_{3,v} \tag{9}$$

Outlet;

$$xe_{8,l} = (h_{8,l} - h_o) - T_o(s_{8,l} - s_o) + \frac{v_{8,l}^2}{2} + gz_{8,l} \tag{10}$$

The general exergy equation can be written as:

$$xE_{r,p} = \dot{m}_p \cdot xe_{r,p} \tag{11}$$

Overall exergy destruction is given by:

$$xE_d = {}_x E_{1,l} + {}_x E_{3,v} - {}_x E_{8,l} \tag{12}$$

It is shown that greater exergy is destroyed with a larger secondary flow inside the system. Figure 18 shows the exergy destruction within the eductor. Future optimization of eductors will include increasing secondary flow into the system as a primary focus; hence, it will also be essential to work toward minimizing exergy destruction to offset this trend and ensure high performance.

The efficiency and exergy destruction at BP_6 is estimated to be minimal. At point BP_6, the back pressure is the highest and the driving pressure difference is the lowest. This results in a minimum secondary flow, and from Equation (7), it can be seen that the efficiency will decrease with a drop in the secondary flow rate; hence the lowest efficiency among the six back pressure conditions tested. Another observation is that at BP_6, the exergy destruction is also minimum, and this also can be attributed to the minimum secondary flow rate. Consider Equation (12), which has three exergy terms, the first and third term in that equation is related to the liquid phase of the fluid and this phase does not have a significant temperature change. The second term is related to the vapor phase, which undergoes a phase change and hence the largest contribution to the exergy. Therefore, in BP_6, the rate of exergy destruction is minimal, corresponding to the lowest mass flow rate of the secondary fluid.

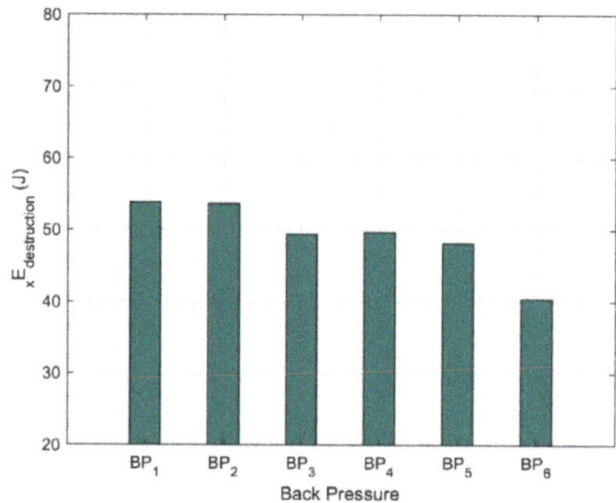

Figure 18. Exergy destruction in Eductor.

5. Eductor for Membrane Distillation

The thermal desalination technologies are energy-intensive processes, but utilization of waste heat (especially low-grade heat) would make it feasible on synergetic (combined recovery of energy and water production) grounds. All thermal systems require large condensers/heat exchangers either for vapor condensation or cooling of the condensing fluid. Although they have large recovery ratios, effective recovery using low-grade heat requires sub-atmospheric vapor generation and large condensers for sensible cooling. The application of eductors for combined vacuum generation, vapor transfer, and condensation would significantly reduce the footprint of existing technology. Eductors are simple static mechanical devices capable of performing multiple fluid functions: pumping, degassing, mixing, direct contact condensing, and compressing. Focusing on MD, replacing the vacuum pump and condenser with an eductor would simplify the existing system and could enhance performance.

Membrane Distillation (MD) is a hybrid technology that is driven thermally and separated through hydrophobic membranes. Based on the permeate flux management method, they are classified into Direct Contact MD (DCMD), Vacuum MD (VMD), Air Gap MD (AGMD), and Sweep Gas MD (SGMD), as shown in Figure 19 [20,21]. All of these processes have a feed and a permeate chamber. At the feed side, continuous feed circulation/recirculation occurs, allowing evaporation at the membrane surface. The hydrophobic nature of the membrane material allows the vapor to cross the membrane boundary while preventing the flow of liquid feed water from crossing. The mechanism of transmembrane vapor transfer mechanism and the management of this vapor differentiate these processes.

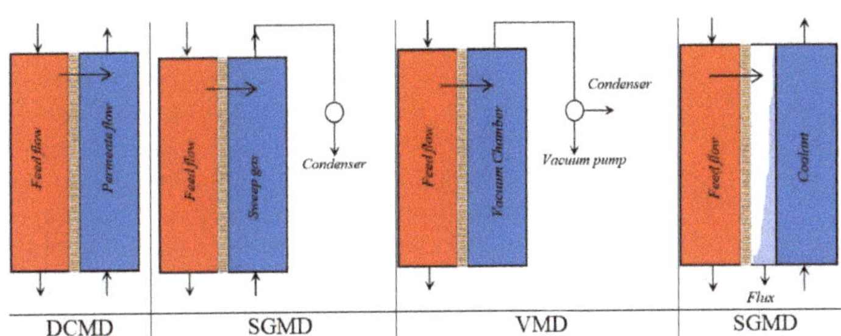

Figure 19. Membrane Distillation Desalination Processes.

For a general Vacuum Membrane Distillation process, feed pressure and temperature, permeate side pressure, permeate vapor suction, condensation, and freshwater transfer are the primary operations and variables. The eductor is capable of vacuum generation, pumping, mixing, heat and mass transfer, and higher-pressure discharge. These capabilities align perfectly with the permeate side operational requirements of a VMD. Figure 20 is the schematic layout of an ejector-based VMD process. The vapor separates from the feed through a hydrophobic membrane (details on VMD can be accessed through), since the permeate chamber pressure is maintained lower than the vapor pressure, condensation does not occur, and the vapor is entrained by the eductor (where there is already primary fluid flow). The primary fluid and secondary fluid (the vapor) mix, transferring mass followed and heat before finally discharging as a single phase fluid at a pressure higher than the secondary (vapor) pressure. Particularly, in an off-grid small-scale desalination unit, this technology could ensure simplicity and reliability. There is great potential for an eductor to improve the sustainable production of freshwater using the abundance of low-grade thermal energy available (solar, industrial waste heat, etc.).

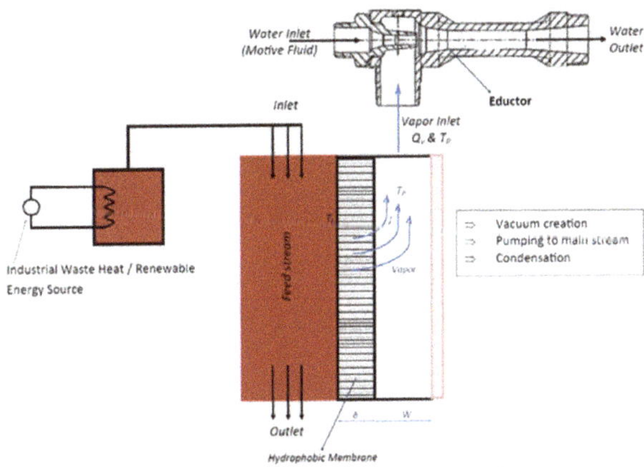

Figure 20. Eductor-based MD process.

6. Conclusions

In contrast to the previous literature, this study developed the eductor as an active vapor transfer and condensation device. A single species fluid (water) with two phases (liquid water as the primary, and water vapor as the secondary) has been used for this study. The prime objective of this application of the eductor is to use the eductor in MD. The study also discusses the operational characteristics of two-phase single species flow within an eductor, to identify the operational range in the developed system.

The series of experiments studied eductor behavior at six different back pressure conditions, and the results have been discussed. With increasing back pressure, the pressure around the suction chamber and the mixing chamber increased. This negatively affected the entrainment of the secondary fluid in the system. One way to identify the zone of condensation was to measure the static pressure. During the phase change from saturated vapor to mixture, a large quantity of vapor bubbles collapse, and the surrounding water molecules rush into the voids. This forms large pressure fluctuations within the flow region, which could be measured. The visualization experiments performed suggest that this filling of voids by liquid is an oscillating phenomenon that is a result of the eductor geometry and plays a vital role during sub-atmospheric thermal saturation.

In applications where the mixing of primary and secondary fluid does not have any influence on the desired product, eductors can be used to maintain sub-atmospheric pressure, secondary pumping, control of vapor generation, and direct contact condensation. The control of all these parameters can be conducted precisely and non-invasively by controlling the back pressure or pressure difference. This strengthens the usability of the eductor in MD for fresh water production.

It was shown that the larger the secondary mass flow rate, the larger the exergy destruction calculated. Hence, design optimization focusing on the entrainment ratio should also target the pressure ratio to have minimum exergy destruction.

This study contributes to flow phenomena studies, along with providing experimental data for verification of computational models related to thermal saturation, mixing of condensing flow, mixing of non-condensing flow, and phase transformation. More studies are required to focus on the factors that influence the frequency of oscillation during direct contact condensation between two-phase flows. In addition to the study of fundamental flow physics, the performance study of eductor-based MD is an additional important topic for the future study.

Author Contributions: Conceptualization, methodology, experiments, writing, R.K., K.I. and A.D.; experiments, visualization, editing, R.K., X.Z., E.R. and A.D.; editing, supervision, K.I. and A.D. All authors have read and agreed to the published version of the manuscript.

Funding: This research received no external funding.

Conflicts of Interest: The authors declare no conflict of interest.

Abbreviations

The following abbreviations are used in this manuscript:

E_r	Entrainment ratio
P_r	Pressure Ratio
$\dot{m}$	Mass Flow rate
ρ	Density
P	Pressure
V	Velocity
T	Temperature
h_c	Heat Transfer Coefficient
BP	Back Pressure
DCC	Direct Contact Condensation
$\eta_{eductor}$	Efficiency of eductor
$_xe$	Specific Exergy
$_xE$	Exergy
Subscript	
S	Suction
i	Inlet
O	Outlet
V	Water Vapor
l	Water Liquid
r	Region
p	Phase
d	Destruction

References

1. Koirala, R.; Ve, Q.L.; Date, A.; Inthavong, I.; Akbarzadeh, A. Influence of inlet pressure and geometric variations on the applicability of Eductor in low temperature thermal desalinations. *J. King Saud Univ. Eng. Sci.* **2021**. [CrossRef]
2. Kumar, R.A.; Rajesh, G. Physics of vacuum generation in zero-secondary flow ejectors. *Phys. Fluids* **2018**, *30*, 066102. [CrossRef]
3. Yang, X.; Long, X.; Yao, X. Numerical investigation on the mixing process in a steam ejector with different nozzle structures. *Int. J. Therm. Sci.* **2012**, *56*, 95–106. [CrossRef]
4. Ksenofontov, B.; Vasilieva, Y.; Kaptinova, S. An ejector for mixing a reagent with discharge water. In *IOP Conference Series: Materials Science and Engineering*; IOP Publishing: Bristol, UK, 2019; Volume 492.
5. Mueller, N.H. Water Jet Pump. *J. Hydraul. Div.* **1964**, *90*, 83–113. [CrossRef]
6. Elbel, S.; Hrnjak, P. Ejector Refrigeration: An overview of Historical and present developments with an Emphasis on Air-Conditioning Applications. In Proceedings of the International Refrigeration Air-conditioning Conference, West Lafayette, IN, USA, 14–17 July 2008.
7. Zhang, Z.; Chong, D.; Yan, J. Modeling and experimental investigation on water-driven steam injector for waste heat recovery. *Appl. Therm. Eng.* **2012**, *40*, 189–197. [CrossRef]
8. Shah, A.; Chughtai, I.R.; Inayat, H. Experimental and numerical analysis of steam jet pump. *Int. J. Mutiphase Flow* **2012**, *37*, 1305–1314. [CrossRef]
9. Banu, J.P.; Mallikarjuna, J.M.; Mani, A. Experimental and numerical investigation of Ejector jet refrigeration system with primary steam swirl. In Proceedings of the International Refrigeration and Air Conditioning Conference, West Lafayette, IN, USA, 11–14 July 2016.
10. Yan, J.; Chong, D.; Wu, X. Effect of swirling vanes on performance of steam–water jet injector. *Appl. Therm. Eng.* **2010**, *30*, 623–630. [CrossRef]
11. Cramers, P.H.; Beenackers, A.A. Influence of the ejector configuration, scale and the gas density on the mass transfer characteristics of gas-liquid ejectors. *Chem. Eng. J.* **2001**, *82*, 131–141. [CrossRef]
12. Cramers, P.H.; Smit, L.; Leuteritz, G.M.; Dierendonck, L.L.V.; Beenackers, A.A. Hydrodynamics and local mass transfer characteristics of gas-liquid ejectors. *Chem. Eng. J. Biochem. Eng. J.* **1993**, *53*, 67–73. [CrossRef]
13. Narabayashi, T.; Mizumachib, W.; Mori, M. Study on two-phase flow dynamics in steam injectors. *Nucl. Eng. Des.* **1997**, *175*, 147–156. [CrossRef]
14. Yan, J.J.; Shao, S.-P.; Liu, J.-P.; Zhang, Z. Experiment and analysis on performance of steam-driven jet injector for district-heating system. *Appl. Therm. Eng.* **2005**, *25*, 1153–1167. [CrossRef]
15. Koirala, R.; Date, A.; Inthavong, K. Numerical study of flow inside water jet eductor. *Exp. Comput. Multiph. Flow* **2021**.
16. Yang, X.; Chong, D.; Liu, J.; Zong, X. Pressure oscillation induced by steam jet condensation in subcooled water flow in a channel. *Int. J. Heat Mass Transf.* **2016**, *98*, 426–437. [CrossRef]
17. Innings, F.; Hamberg, L. Steam condensation dynamics in annular gap and multi-hole steam injectors. *Procedia Food Sci.* **2011**, *1*, 1278–1284. [CrossRef]
18. Kim, H.J.; Lee, S.C.; Bankoff, S.G. Heat transfer and interfacial drag in countercurrent steam-water stratified flow. *Int. J. Multiphase Flow* **1985**, *11*, 593–606. [CrossRef]
19. Cengel, Y.A.; Boles, M.A. *Thermodynamics: An Engineering Approach*; McGraw-Hill Education: New York, NY, USA, 2014; Volume 8.
20. Belessiotis, V.; Kalogirou, S.; Delyannis, E. Membrane Distillation. In *Thermal Solar Desalination: Methods Systems*; Elsevier: Amsterdam, The Netherlands, 2016; pp. 191–251.
21. Lawson, K.W.; Lloyd, D.R. Membrane Distillation. *J. Membr. Sci.* **1997**, *124*, 1–25. [CrossRef]

 *energies*

Article

Solar Energy Driven Membrane Desalination: Experimental Heat Transfer Analysis

Hosam Faqeha [1,2,*], Mohammed Bawahab [2,3], Quoc Linh Ve [2,4], Oranit Traisak [2], Ravi Koirala [2], Aliakbar Akbarzadeh [2] and Abhijit Date [2,*]

[1] Mechanical Engineering Department, College of Engineering and Islamic Architecture, Umm Al-Qura University, P.O. Box 715, Makkah 21955, Saudi Arabia

[2] Mechanical and Automotive Engineering, School of Engineering, RMIT University, Bundoora, VIC 3083, Australia

[3] Department of Mechanical Engineering, Faculty of Engineering, University of Jeddah, Jeddah 23890, Saudi Arabia

[4] Faculty of Engineering and Food Technology, University of Agriculture and Forestry, Hue University, Hue 530000, Thua Thien Hue, Vietnam

[*] Correspondence: hzfaqeha@uqu.edu.sa (H.F.); abhijit.date@rmit.edu.au (A.D.)

Abstract: In the direct contact membrane distillation (DCMD) system, the temperature polarization due to boundary layer formation limits the system performance. This study presents the experimental results and heat transfer analysis of a DCMD module coupled with a salinity gradient solar pond (SGSP) under three different flow channel configurations. In the first case, the feed and permeate channels were both empty, while in the next two cases, the feed and permeate channels were filled with a porous spacer material. Two different spacer geometries are examined: 1.5 mm thick with a filament angle of 65°, and 2 mm thick with a filament angle of 90°. The study considers only the heat transfer due to conduction by replacing the hydrophobic membrane normally used in a DCMD module with a thin polypropylene sheet so that no mass transfer can occur between the feed and permeate channels. The Reynolds number for all three configurations was found to be between 1000 and 2000, indicating the flow regime was laminar. The flow rate through both the feed and permeate sides was the same, and experiments were conducted for flow rates of 5 L/min and 3 L/min. It has been found that the highest overall heat transfer coefficient was obtained with the spacer of 2 mm thickness and filament angle of 90°.

Keywords: water desalination; membrane desalination; solar pond; heat transfer

Citation: Faqeha, H.; Bawahab, M.; Linh Ve, Q.; Traisak, O.; Koirala, R.; Akbarzadeh, A.; Date, A. Solar Energy Driven Membrane Desalination: Experimental Heat Transfer Analysis. *Energies* **2022**, *15*, 8051. https://doi.org/10.3390/en15218051

Academic Editor: Alessandra Criscuoli

Received: 15 October 2022
Accepted: 25 October 2022
Published: 29 October 2022

Publisher's Note: MDPI stays neutral with regard to jurisdictional claims in published maps and institutional affiliations.

1. Introduction

Water is a common, widely available substance. However, only about 3% of it is available for human consumption. The remaining 97% is seawater. The availability of even this 3% is affected by various factors. Environmental pollution is one problem. According to the World Health Organization (WHO) and the United Nations International Children's Emergency Fund (UNICEF), about 2.1 billion people are deprived of access to safe, readily available water at home and about 4.5 billion people do not have safe sanitation facilities [1]. The problem is severe in the rural and remotely located areas of developing countries. Although access to water has increased over the years, access to safe water and sanitation is still a severe problem. The global population was 7.66 billion in November 2018 and is expected to reach 8.5 billion by 2030 and 9.7 billion by 2050 [2]. It has been increasing rapidly since the industrial revolution of the 1800s, growing from 1.65 billion to 6 billion during the 20th century. However, it is expected to take about 200 years for the current population to double, compared to the 58 years it took to double from three billion to six billion. However, the population is expected to increase at a much higher rate in certain regions such as the Middle East and Sub-Saharan Africa, South Asia,

Southeast Asia, and Latin America. Such growing population density in certain areas can lead to heavy water demand combined with more significant pollution problems.

Traditional water treatment methods such as multistage flash (MSF), multi effect distillation (MED), and reverse osmosis (RO) need either high temperature thermal energy or electrical energy for operation. On the other hand, membrane desalination can operate at much lower temperatures and hence in recent years there has been a significant research focus on the development of membrane desalination systems. In this process, a microporous hydrophobic membrane is used to separate liquids from dissolved solids [3]. The general types of MD system configurations listed below are shown in Figure 1:

- direct contact membrane distillation (DCMD);
- sweep gas membrane distillation (SGMD);
- vacuum membrane distillation (VMD);
- air gap membrane distillation (AGMD).

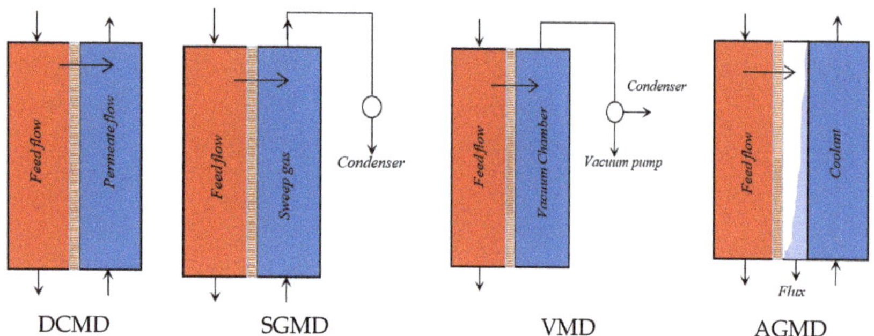

Figure 1. Schematic of different types of membrane desalination systems.

In DCMD, the feed solution is heated and is in direct contact with the surface of the membrane. Therefore, quick evaporation occurs at the feedwater–membrane interface. Vapor moves across the membrane due to the vapor pressure difference and then condenses inside the membrane module. The hydrophobic nature of the membrane prevents the liquid feedwater from penetrating the membrane, hence, vapor can only exist inside the membrane pores. The DCMD configuration is the most straightforward configuration of membrane distillation technologies. Hence, there is more widespread use of DCMD compared to these other configurations. The heat lost by conduction is a significant limitation of DCMD. Although the thermal conductivity of polymeric membranes is typically minimal, the driving force for the desalination, i.e., the temperature differential, results in considerable conductive heat transfer through the membrane material due to its minimal thickness. As a result, DCMD has the highest energy loss by thermal conduction of any MD configuration, resulting in low thermal efficiency [4]. DCMD is used in a variety of industrial applications, including the desalination and concentration of aqueous solutions in food industries and the manufacture of acids [5,6].

DCMD can be used for desalination and concentration of aqueous solutions; however low energy efficiency is a major drawback affecting its use in large-scale applications. The mass flux and energy efficiency can be increased by reducing the temperature polarization. The first step for this is to understand the effects of flow regime on the heat transfer in the absence of mass transfer. Hence, the estimation of heat transfer coefficients is essential for analysis and design of efficient MD modules. Some recent research work on these aspects is reviewed below.

An early study on the use of spacer-filled channels in MD applications by Martınez-Dıez, Vázquez-González, and Florido-Dıaz [7] showed advantages in terms of enhanced heat and mass transfers using low-grade heat sources. In commercially available membrane distillation modules, the feed and permeate flow channels have different spacer material and orientation. In a numerical study, Taamneh [8] found that the spacer increased shear stress at the wall and doubled the Nusselt number in contrast to an empty channel. In a DCMD module, mass flux enhancement by spacers was detected by Phattaranawik, Jiraratananon, Fane, and Halim [9]. The authors developed a model to explain mass flux increases due to spacers. Phattaranawik et al. [10] tested net type spacers fitted in DCMD to enhance mass transfer coefficient. In their work, Yun, Wang, Ma, and Fane [11] studied the effects of spacers on flux enhancement of DCMD using a high concentration NaCl aqueous solution. The observed increase in mass flux was highest for the thick spacer, then thin spacer, and then without spacer. Taamneh and Bataineh [12] used experimental and numerical methods to test the effect of presence and orientation of filaments in spacers. An empty channel was used for comparison.

A study on the trans-membrane heat and mass transfer using comprehensive 3D CFD simulation covering the entire length of the module done by Chang, Hsu, Chang, and Ho [13] showed that spacers created high velocity regions near the membrane surfaces. The reputation in spacer geometry results in irregularity in heat and mass flux.

Gong et al. [14] proposed a new design with solar energy and graphene membrane, called solar vapor gap membrane distillation (SVGMD). They showed that this design has high energy efficiency and long-term stability and anti-fouling properties.

Quoc Linh Ve et al. [15] performed experimental analysis to determine the coefficient of heat and mass transfer for DCMD. The experiment used copper plate for different conditions; empty and spacer filled. The heat transfer correlations were within an acceptable limit.

In a different application of MD, AGMD, Chernyshov, Meindersma, and De Haan [7] investigated five geometries with the same thickness and different geometry. With spacers, about 2.5 times higher flux was noted when compared with an empty channel, in another application of MD

This seems to be higher than noted for DCMD. Different types of spacer configurations were determined to be optimum depending on the different levels of importance attached to either temperature or mass fluxes.

The objective of this experimental work is to determine the overall heat transfer coefficient of a DCMD module with and without a spacer-filled channel to help understand the effects of spacer on the temperature polarization. The channel gap of the DCMD module was fixed, and the tests were conducted for three different configurations; empty channel, thin spacer-filled channel, and thick spacer filled channel and compared with heat transfer correlations presented in the literature to find the best fit.

2. Theory on Heat Transfer Phenomena in DCMD

Complex transport processes, including mass and heat transfer, occur concurrently throughout the DCMD process as seen in Figure 2. A DCMD module is typically composed of a flat module which has a feed chamber (for hot feedwater) and a cooling permeate chamber separated by a hydrophobic microporous membrane. Heat transfer occurs via convection/conduction across membrane (from feed to permeate) and convection/mass transfer (vapor transport) through the membrane pores.

On the feed side, the feedwater temperature (T_f) drops from the bulk fluid temperature to the membrane surface temperature (T_{mf}) across the boundary layer. As the vapor from the feed side condenses in the water on the permeate side, the permeate stream temperature increases. This results in a decreasing temperature gradient between the permeate fluid nearest the membrane (T_{mp}), through the boundary layer to the bulk fluid stream (T_p). The driving force is the difference in vapor pressure resulting from the difference in interface temperatures on the feed and permeate sides (T_{mf} and T_{mp}). This is lower than the difference between bulk feed and permeate temperatures (T_f and T_p).

The temperatures near the membrane surfaces vary from the bulk fluid temperatures due to the heat transfer that occurs throughout the DCMD process. This results in the reduction in the driving force, and hence mass flux, compared to what would be expected based on the bulk fluid temperatures. This "is known as temperature polarization and the temperature polarization coefficient (*TPC*) is the ratio between the actual driving force and the theoretical driving force" [16–18].

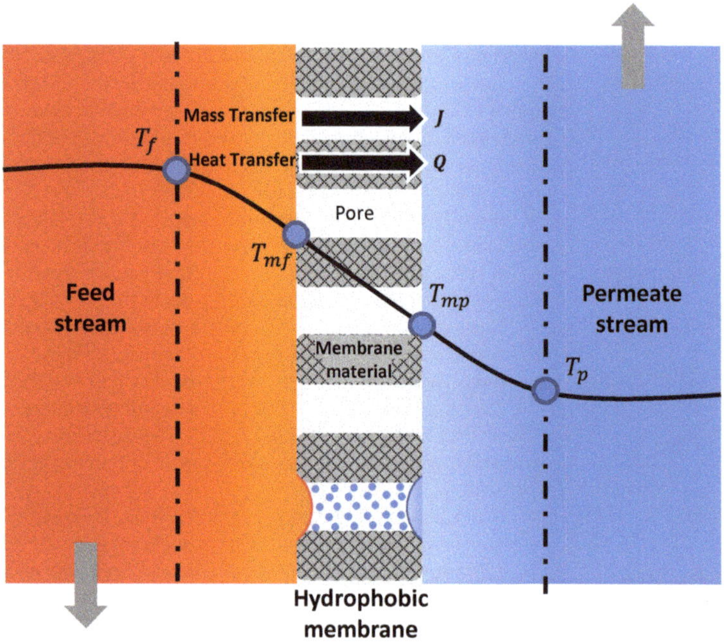

Figure 2. Schematic diagram of the transport mechanism for the direct contact membrane distillation (DCMD) process.

The Laplace–Young equation describes the pressure differential between liquid–vapor interfaces. The liquid entry pressure (*LEP*) of a membrane is defined as the lowest possible value of the hydrostatic pressure difference at which the feed liquid may pass through the biggest holes of the membrane. The interfacial tension, the liquid's contact angle at the pore entrance, and the size and shape of membrane pores all have a role in the liquid entry pressure. Franken et al. proposed the simple approach for determining the *LEP* value using the Laplace–Young equation [19].

Although they are not specifically mentioned, the operating temperature and the composition of the process solution may have a considerable effect on the liquid–solid contact angle and liquid surface tension. Therefore, when choosing a membrane, these impacts should not be ignored.

PTFE, PP, and polyvinylidene fluoride (PVDF) are the most often utilized materials for MD membranes. MD membrane porosity has been observed to vary between 35% and 93%, pore size typically varies between 100 nm and 1 μm, and membrane thickness typically s between 0.04 and 0.25 mm [20].

2.1. Mechanism of Heat Tramsfer

Heat transfer in the proposed module configuration with impermeable membrane occurs in three regions, defined in Figure 3. Heat transfer by convection occurs on the feed side from the fluid to the membrane; heat is then transferred by conduction through the thin plastic sheet; and finally, heat transfer by convection occurs on the permeate side from the membrane to the fluid. It should be noted that heat transfer resulting from the mass transfer would also occur if a vapor-permeable membrane was used. Figure 3 depicts the heat transfer thermal resistance model used in this study.

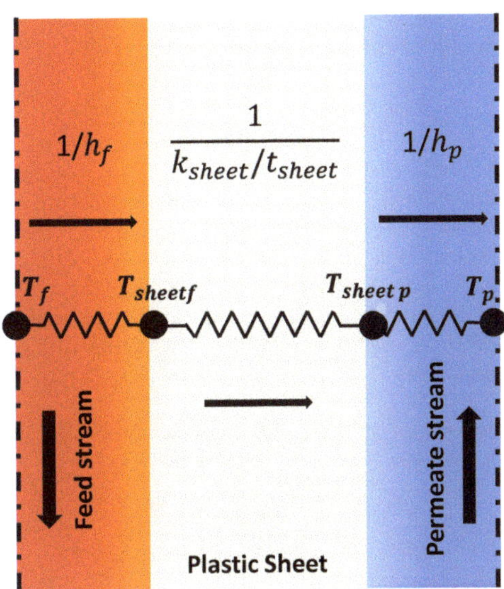

Figure 3. The thermal resistance series in empty and spacer-filled channels.

Equations (1) and (2) give convective heat transfer across feed to permeate side [21]:

$$\dot{Q}_f = \dot{m} \times C_p \times \left(T_{f,in} - T_{f,out} \right) \tag{1}$$

$$\dot{Q}_p = \dot{m} \times C_p \times \left(T_{p,out} - T_{p,in} \right) \tag{2}$$

$$\dot{Q}_f = h_f \times A_m \times \left(T_f - T_{sheet,f} \right) \tag{3}$$

$$\dot{Q}_p = h_p \times A_m \times \left(T_{sheet,p} - T_p \right) \tag{4}$$

where the difference between $T_{sheet,f}$ and $T_{sheet,p}$ is assumed to be negligible due to the small thickness of the plastic sheet (0.1 mm).

Here T_f and T_p; are the average bulk temperature of fluid for both sides. It is calculated using Equations (5) and (6)

$$T_f = \frac{T_{f,in} + T_{f,out}}{2} \tag{5}$$

$$T_p = \frac{T_{p,out} + T_{p,in}}{2} \tag{6}$$

The following Equation (7) can be used to determine the overall heat transfer coefficient of the DCMD module:

$$U = \left[\frac{1}{h_f} + \frac{t_{sheet}}{k_{sheet}} + \frac{1}{h_p} \right]^{-1} \tag{7}$$

2.2. Empty Channel

For the case where an empty channel is used, the experiments were conducted using the DCMD configuration module. The heat transfer coefficients are assumed to equal the same mass flow rate for both feed and permeate sides, and channel geometry was used. The Nusselt number for the experiments can be found in Equation (8)

$$Nu_{exp} = \frac{h * D}{k} \tag{8}$$

where $h = h_f = h_p$, and k and D are thermal conductivity and hydraulic diameter, respectively. Hydraulic diameter D can be determined by

$$D = \frac{2 * W * t}{W + t} \tag{9}$$

Table 1 shows different heat transfer correlations suggested in the literature. In order to determine the optimal heat transfer correlation for the DCMD configurations considered here, these correlations are used to determine a theoretical Nusselt number. These are then compared to the Nusselt numbers obtained from the experiments discussed in Sections 3 and 4. The best correlation is based on the minimum variation between experimental and theoretical Nusselt numbers, for a given configuration. The Reynolds number and Prandtl number required for the correlations in Table 1 can be estimated using the following equations.

Table 1. Different types of heat transfer correlation with Reynolds number in laminar flow range.

Correlation	Equation Number	Reference
$Nu = 1.86 \left(\dfrac{Re\,Pr}{L/D} \right)^{\frac{1}{3}}$	(10)	[22]
$Nu = 1.95 \left(\dfrac{Re\,Pr}{L/D} \right)^{\frac{1}{3}}$	(11)	[23]
$Nu = 4.36 + \dfrac{0.036 Re\,Pr(D/L)}{1 + 0.0011(Re\,Pr(D/L))^{0.8}}$	(12)	[21]
$Nu_{cooling} = 11.5(Re\,Pr)^{0.23}(D/L)^{0.5},$ $Nu_{heating} = 15(Re\,Pr)^{0.23}(D/L)^{0.5}$	(13)	[24]
$Nu = 0.13 Re^{0.64} Pr^{0.38}$	(14)	[24]
$Nu = 0.097 Re^{73} Pr^{0.13}$	(15)	[25]
$Nu = 3.66 + \dfrac{0.104 Re\,Pr(D/L)}{1 + 0.0106(Re\,Pr(D/L))^{0.8}}$	(16)	[23]
$Nu = 4.86 + \dfrac{0.06063(Re\,Pr(D/L))^{1/2}}{1 + 0.09094(Re\,Pr(D/L))^{0.7} Pr^{0.17}}$	(17)	[26]
$Nu = 1.62 \left(\dfrac{Re\,Pr}{L/D} \right)^{0.33}$	(18)	[21]

$$Re = \frac{\rho * u * D}{\mu} \tag{19}$$

$$Pr = \frac{\mu * C_p}{k} \tag{20}$$

where u is the superficial velocity of the fluid flow and can be determined by,

$$u = \frac{V}{W * t} \tag{21}$$

the overall heat transfer coefficient can be determined from the experimental results using Equations (22) and (23) as suggested by Ve and Rahaoui [15].

$$U = \frac{\dot{Q}}{A * \Delta T_{LMTD}} \tag{22}$$

where:

$$\Delta T_{LMTD} = \frac{\left(T_{f,in} - T_{p,out}\right) - \left(T_{f,out} - T_{p,in}\right)}{ln\left[\left(T_{f,in} - T_{p,out}\right)/\left(T_{f,out} - T_{p,in}\right)\right]} \tag{23}$$

2.3. Spacer-Filled Condition

The net-type spaces are common in membrane modules for commercial systems (ultrafiltration/spiral wound reverse osmosis). They can provide structural support to the membrane and, depending on the orientation of the spacer, can also cause the fluid flow to transition from a laminar to a localized turbulent flow regime. In a DCMD module, turbulent flow improves heat transfer. This increases the driving temperature difference, and hence the production of freshwater, and decreases the effect of temperature polarization. The spacer orientation, geometry, and its relation to the flow direction are illustrated in Figure 4.

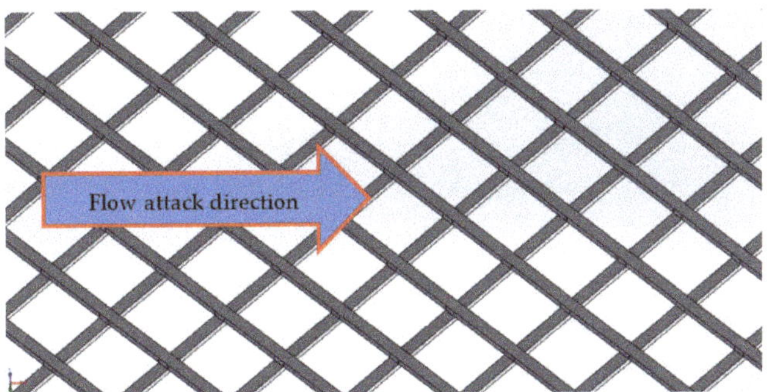

Figure 4. Spacer used to fill DCMD module's channels including orientation, geometry, and flow attack angle.

For spacers that influence the fluid flow:

$$Nu^s = 0.664 * k_{dc} * Re^{0.5} * Pr^{0.33} * \left(\frac{2 * d_{hs}}{l_m}\right)^{0.5} \tag{24}$$

and

$$k_{dc} = 1.654 * \left(\frac{d_f}{t_s}\right)^{-0.039} \varepsilon^{0.75} * \left(sin\left(\frac{\theta}{2}\right)\right)^{0.086} \tag{25}$$

where d_{hs} is hydraulic diameter for spacer filled condition, k_{dc} is spacer geometry geometry correction factor, l_m is the mesh size, H is the spacer thickness, ε is the spacer voidage, and θ the hydrodynamic angle.

For spacers that do not influence the flow direction (one set of filaments is parallel to, and the other is transverse to the flow direction):

$$Nu^s = 0.664 * k_{dc} * Re^{0.5} * Pr^{u} * \left(\frac{d_{hs}}{l_m}\right)^{0.5} \tag{26}$$

where Re and Pr are calculated by using Equations (19) and (20), respectively, and u is 0.33. The following equation can determine the superficial velocity for a spacer filled channel [27]:

$$u = \frac{V}{W * t * \varepsilon} \tag{27}$$

where ε is the spacer voidage which is defined by the following equation [28]:

$$\varepsilon = 1 - \frac{\pi * d_f^2}{2 * l_m * t_s * sin\theta} \tag{28}$$

d_{hs} is the spacer-filled channel hydraulic diameter and is defined by the following equation [29]:

$$d_{hs} = \frac{4 * \varepsilon}{\left(\frac{2*(W+t)}{W\,t}\right) + (1 - \varepsilon) * S_{vsp}} \tag{29}$$

The specific surface of the spacer is defined as

$$S_{vsp} = \frac{4}{d_f} \tag{30}$$

An alternative Nusselt number correlation for the case of a spacer- filled channel has been suggested by Schwager and Robertson [30] and is given by Equation (30):

$$Nu^s = 1.38 * Re^{0.483} * Pr^{0.33} * \left(\frac{d_{hs}}{l_m}\right)^{0.531} \tag{31}$$

Laminar and turbulent condition in channel flow and its influence on heat transfer coefficient can be calculated using Equation (32) suggested by Zhang and Gray [31]

$$Nu^s = k_{dc} * 0.023 * \left[1 + 6 * \left(\frac{d_{hs}}{L}\right)\right] Re^{0.8} * Pr^{0.33} \tag{32}$$

where:

$$k_{dc} = 1.923 * \left(\frac{d_f}{t_s}\right)^{-0.168} |sin\theta|^{0.292} \; exp\left[-1.601 * \left|ln\left(\frac{\varepsilon_{sp}}{0.6}\right)^2\right|\right] \tag{33}$$

Phattaranawik [10] proposed another correlation for space filled case. It can be determined from Equation (34) below:

$$Nu = k_{dc}\left[4.36 + \frac{0.036 * Re * Pr * (D/L)}{1 + 0.0011 * (Re * Pr * (D/L))^{0.8}}\right] \tag{34}$$

3. Experimental Setup

A schematic of the experimental setup for the DCMD module in the Renewable Energy Lab at RMIT University is shown in Figure 5. The solar pond is made of three layers, namely, lower convective zone (LCZ), non-convective zone (NCZ) and upper convective zone (UCZ). The DCMD module contains feed and permeate channels, normally separated by a hydrophobic membrane. For these experiments, a thin plastic sheet (clear polypropylene of 100 μm thickness) replaces the hydrophobic membrane to exclude mass transfer between the feed and permeate channels. Freshwater is circulated through both feed and permeate channels as a heat transfer fluid. A 24 V, self-priming diaphragm pump with a maximum flow rate of 8 L/min is used to pump the water through the system. The feed reservoir tank is filled with 100 L of freshwater, which is pumped to an in-pond heat exchanger in the SGSP to increase the temperature of the feed side fluid. In addition, an evacuated tube solar collector combined with a thermal storage tank (E.T. tank) is used as an auxiliary heat supply. After leaving the SGSP, the freshwater from the feed reservoir was circulated through a heat exchanger in the E.T tank before being fed into the system and returned to the reservoir. In the case of the permeate side, freshwater is transferred from the permeate tank to the top layer of the SGSP to dissipate the heat gained from the feed side fluid. After that, the cooled freshwater passes through the permeate channel in the DCMD module and is then returned to the permeate tank. The 3D structural arrangement of the DCMD module is presented in Figure 6 and shows the channel gaps, spacers, and membrane sheet.

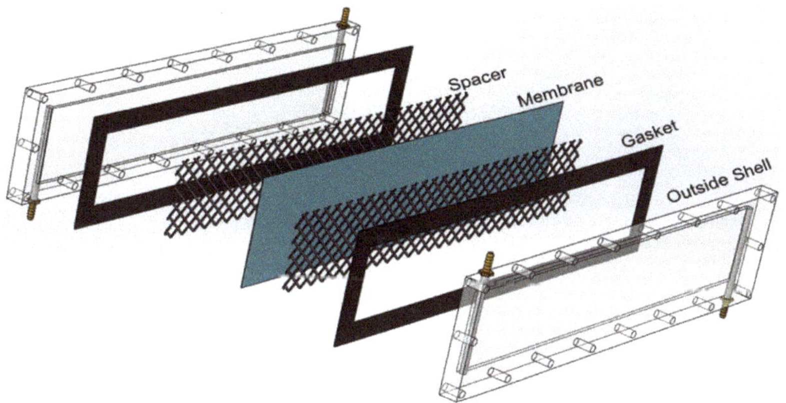

Figure 5. Schematic configuration of the sustainable experimental system using a thin plastic sheet.

Figure 6. The 3D structural arrangement of the DCMD module.

There are two loops in the proposed system: feed loop and permeate loop. Feed loops experience loss of mass and energy while permeate loops gain mass and energy. Applying energy balance to the feed loop which is connected to the solar pond LCZ and external heater, we can estimate the feed water supply temperature to the feed channel as

$$T_{out\_f} = T_{in\_f} + \left[\frac{m_{SGSP\_LCZ} \times c_{lcz\ water}}{\dot{m}_f \times c_{feed}} \times \frac{\Delta T_{lcz}}{\Delta t}\right] + \left[\frac{\dot{Q}_{heater}}{\dot{m}_f \times c_{feed}}\right] \tag{35}$$

Here, T_{out_f} is the feed water temperatue after it is heated in LCZ and external heater. The mass flow rate of the feed $\dot{m}_f$ is constant while it is getting heated in solar pond and external heater. The specific heat capacity of the feed water and the saline water of LCZ is shown as c_{feed} and $c_{lcz\ water}$, respectively. On the left hand side, the second term in the square brackets represents the temperature rise of feed water in the solar pond and the third term represent temperature rise in the external heater. In the second square bracket, the mass of the water in solar pond LCZ is given as m_{SGSP_LCZ}, while the ΔT_{lcz} represents change in the temperature of the solar pond LCZ in Δt time step. The time step is relative to the time rate that is used for mass flow rate. The mass and energy balance in the feed channel can be written as

$$\dot{m}_f \times c_{feed} \times T_{in\_f\_ch} = \left(\dot{m}_f - \dot{m}_{p'}\right) \times c_{feed} \times T_{out\_f\_ch} + \dot{Q}_{cond} + \dot{m}_{p'} \times h_{fg@T_{f\_avg}} \tag{36}$$

Here, the term $\dot{m}_{p'}$ represents the rate of mass transfer that happens from the feed to the permeate side. The temperature of the feed inlet is estimated from Equation (35) and is given as $T_{in\_f\_ch}$. The change in the mass of the feed due to vapor transport to permeate side is given as $\left(\dot{m}_f - \dot{m}_{p'}\right)$, while the heat transfer through the solid parts of the membrane is given as $\dot{Q}_{cond}$. The mass and energy balance in the permeate channel can be written as,

$$\dot{m}_p \times c_{permeate} \times T_{out\_p} = \left(\dot{m}_p + \dot{m}_{p'}\right) \times c_{permeate} \times T_{in\_p} + \dot{Q}_{cond} + \dot{m}_{p'} \times h_{fg@T_{p\_avg}} \tag{37}$$

Here, the term $\dot{m}_p$ represents the mass flow rate of cold permeate that is coming into the permeate channel. The permeate inlet and outlet temperatures are given as T_{out_p} and T_{in_p}. The mass of the fresh permeate that is added to the permeate flow is recovered as an overflow from the permeate tank. The warm permeate is cooled in the solar pond UCZ heat exchanger and, if needed, in an external cooler. Applying energy balance, we can estimate the temperature of the permeate after the cooling T_{out_p1} as

$$T_{out\_p1} = T_{in\_p} - \frac{m_{SGSP\_UCZ} \times c_{Ucz\ water}}{\dot{m}_p \times c_{permeate}} \times \frac{\Delta T_{ucz}}{\Delta t} + \frac{\dot{Q}_{cooler}}{\dot{m}_p \times c_{permeate}} \tag{38}$$

4. Result and Discussion

The following sections cover the experimental results and discussion of the heat transfer within the DCMD module for both empty channel and spacer filled channel conditions. The experimental results are compared with theoretical Nusselt number correlations to determine the most suitable of these to use for further numerical modelling.

4.1. DCMD Heat Transfer with Empty Channels

In the case of empty channels, the experiment ran with the same setup conditions shown in Figure 7 with two different flow rates. First, a counter flow arrangement uses 3 L/m on both feed and permeate sides. Then, it uses 5 L/m in both channels, also with counter flow. Figure 7 shows the overall heat transfer coefficient calculated from the experimental results for the two different flow rates. The average overall heat transfer coefficient for the 3 L/m tests was 593 W/m^2·°C, and for 5 L/m, it was 724 W/m^2·°C. The experimental results show that the overall heat transfer coefficient increases by approximately 18% at steady state when the flow rate increases from 3 to 5 L/min. In Figures 8 and 9, the feed side outlet temperature (T_{f_out}) is lower than the inlet temperature (T_{f_in}) because of the occurrence of heat transfer to the permeate side. Correspondingly, the outlet temperature of the permeate side (T_{p_out}) increases due to heat transfer through the plastic sheet.

The Reynolds numbers for the two experimental conditions were calculated using Equation (19). They suggested that the flow regime was laminar in both cases. The theoretical Nusselt numbers from heat transfer correlations summarized in Table 1 were calculated for the experimental flow rate conditions presented in Figures 10 and 11. The correlation given by Equation (12) was the most appropriate correlation with a deviation of 10% between the theoretical and the experimental overall heat transfer coefficients.

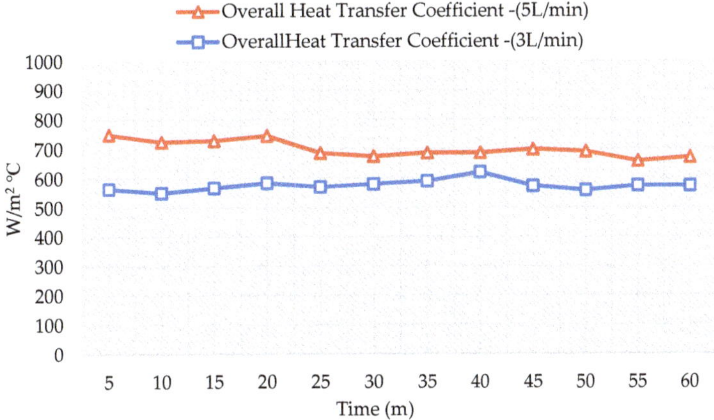

Figure 7. The overall heat transfer coefficient experimentally with different flow rates for the empty channels.

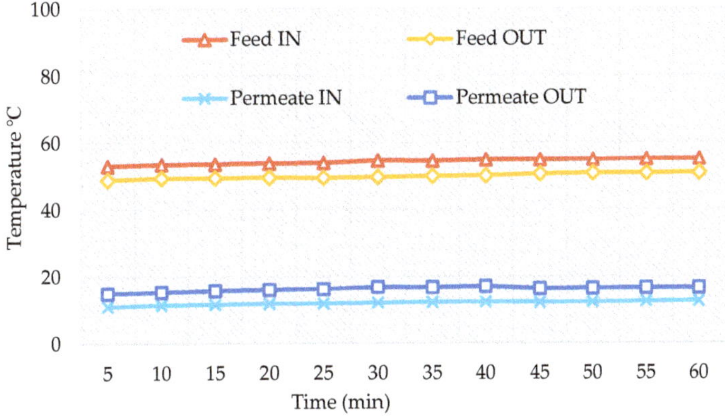

Figure 8. Feed and permeate temperature in case of the empty channel by using 3 L/min.

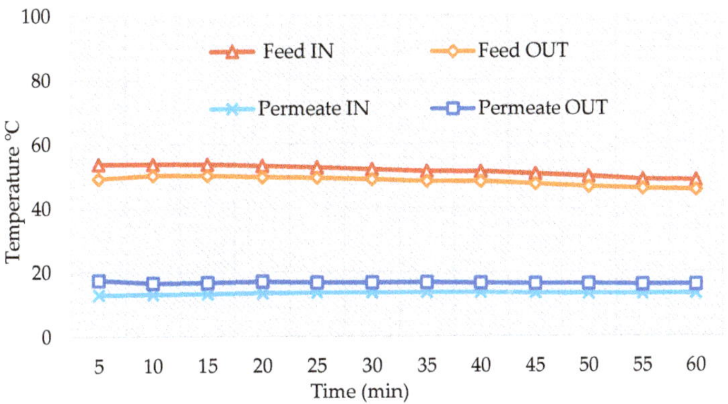

Figure 9. Feed and permeate temperature in case of the empty channel by using 5 L/min.

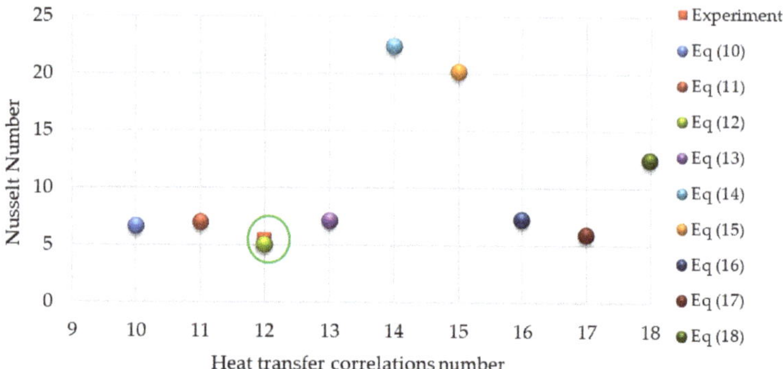

Figure 10. Correlation for empty channel (flow rate = 3 L/m).

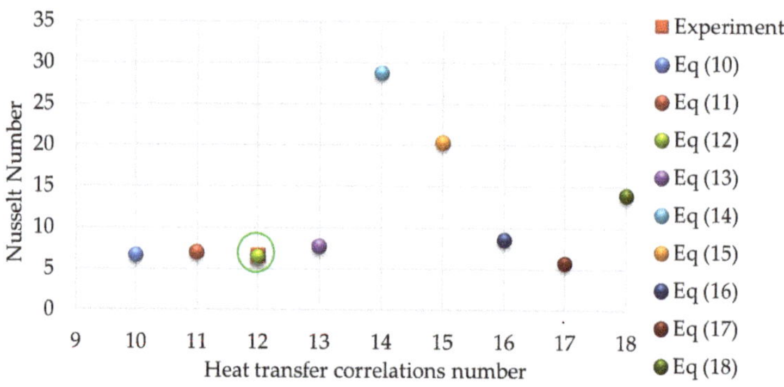

Figure 11. Correlation for empty channel (flow rate = 5 L/m).

4.2. DCMD Spacer-Filled Condition

This system conducted the experiments on the same large-scale DCMD module used for the empty channel experiments with the two different spacer geometries shown in Figure 12 and Table 2. Figure 12(1) shows thinner mesh whereas Figure 12(2) shows thicker mesh.

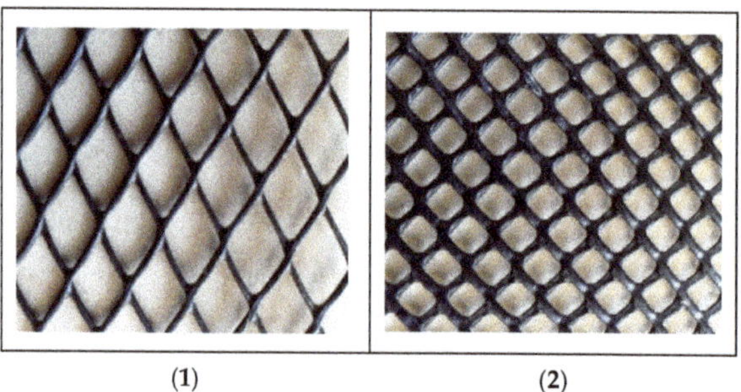

Figure 12. Non-woven plastic spacers (**1**) thin spacer, (**2**) thick spacer.

Similar to the empty, spacer-filled channel, these experiments investigated the conductive heat transfer with each spacer geometry using two flow rates: 3 L/min and then 5 L/min. It can be seen in Figures 13 and 14 that the overall heat transfer coefficient in the case of the spacer-filled channels is higher in the case of empty channels for both 3 and 5 L/min. Comparing just the heat transfer coefficients obtained with each of the two kinds of spacer, it can be seen that the thicker (2 mm) spacer produces a higher heat transfer coefficient. The average values for overall heat transfer in the case of the spacer-filled channel with a 1.5 mm thick spacer and flow rates of 3 and 5 L/min, were 949 W/m^2·°C and 1379 W/m^2·°C, respectively. Furthermore, the average overall heat transfers for a spacer-filled channel with a 2 mm thick spacer and flow rates of 3 L/m and 5 L/m were 1030 W/m^2·°C and 1465 W/m^2·°C, respectively.

Table 2. Characteristics of spacers.

No.	Spacer	Material	Length (m)	Width (m)	Thickness (m)	Filament size, d_f, (m)	Angle, θ, (°)	Mesh Size, l_m, (m)	Porosity (%)
1	Non-woven	Plastic	0.69	0.1339	0.0015	0.0008	66.5	0.0075	90
2	Non-woven	Plastic	0.69	0.1339	0.0020	0.00156	90	0.0044	57

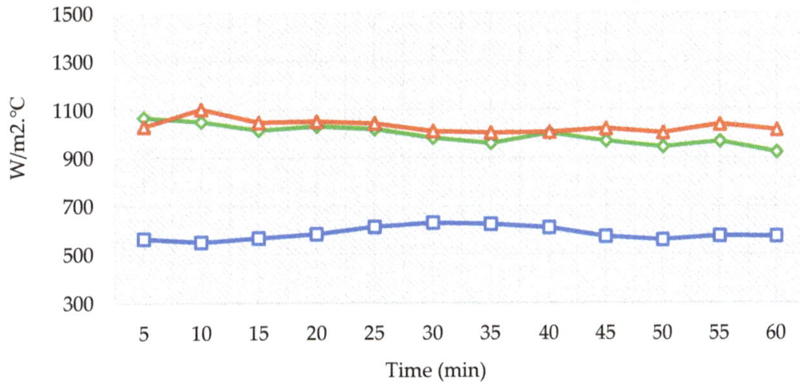

Figure 13. Heat transfer in the empty and spacer-filled channels by using 3 L/min.

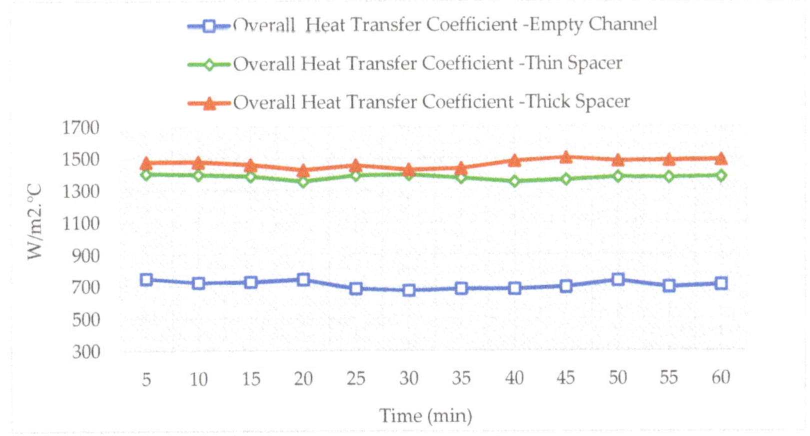

Figure 14. Heat transfer in the empty and spacer-filled channels by using 5 L/min.

It is essential to mention that the DCMD channel thickness was fixed, while the thickness of the two spacers was different. This allowed the thin spacer to float between the plastic sheet (membrane) and the wall of the DCMD module. The DCMD channel thickness was 2.8 mm, whereas the thin spacer thickness was only 1.5 mm. Therefore, around 46% of the channel was free. As a result, the thin spacer could have effectively created a thinner 'empty channel', which might have influenced the mechanism underlying the increase in the overall heat transfer coefficient observed for the thin spacer compared to the thick spacer. Thicker spacers have also been applied in later experiments, where the spacer filled the channel completely. The thicker space has been observed to produce a slightly better heat transfer performance compared to the thinner spacer, but the overall heat transfer results of both thin and think spacer is very close and cannot be used to call any one of the arrangements better than other.

Based on Equation (19), the Reynolds number was lower than 2100 for all of the channel configurations and flow rates investigated. By knowing the Reynolds number, the flow regime can be predicted, with a number between 600 and 2100 indicating a laminar flow. For this experimental investigation, the bulk flow regime could be considered as laminar for both flow rates and spacers used. The addition of the spacer reduces the height of the cavity and increases the aspect ratio, and this increases the number of eddies [32]. These localized eddies break the boundary layer and reduces the temperature polarization. In the present experiments, the membrane surface temperature is not measured, but the addition of spacers has shown to improve heat transfer. Based on this observation, it is better to have channels with high aspect ratio for a MD system.

Many researchers have investigated correlations for different kinds of channels; their findings were applied to find the best agreement with the experimental results presented from Figures 7–14. A summary of all the experimental and theoretical Nusselt numbers is shown in Figures 15–17. Both Phattaranawik and Jiraratananon [33] and Kim and Francis [34] presented heat transfer correlations for a DCMD module with non-woven spacer-filled channels. Using their correlations, Equations (24) and (26) under the current experimental conditions (1.5 mm and 2 mm spacers; and 3 and 5 L/m flow rates) produced Nusselt number values that differed from experimental results by 90%. Correlations are also given by Equations (31) and (32), which overpredicted the Nusselt number values by 55–72% as compared to experimental results.

The smallest deviation was found using Equation (34) for a 1.5 mm thick non-woven spacer. For both flow rates, the deviation was 10–20%. However, the deviation using the same equation for 2 mm thick spacer-filled channels was greater at 42–55%. Therefore, Equation (34) produces the most appropriate heat transfer correlation among all the equations compared above. Nonetheless, the results from this correlation are still not satisfactory for all cases.

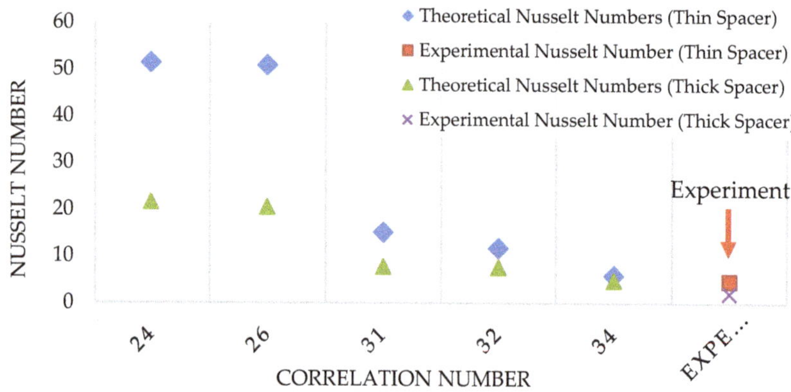

Figure 15. Correlations for spacer-filled channels with flow rate (3 L/min).

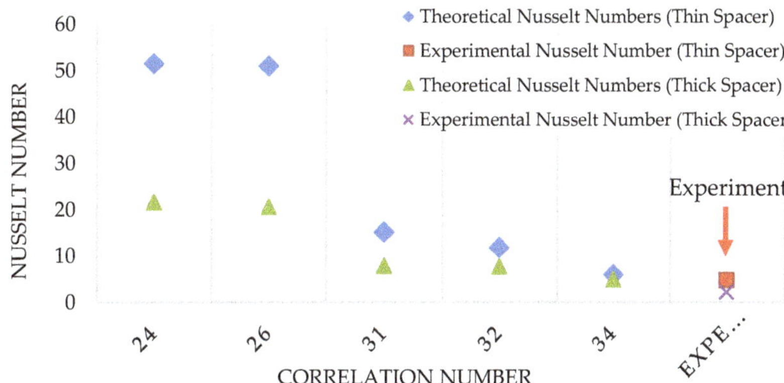

Figure 16. Correlations for spacer-filled channels with flow rate (5 L/min).

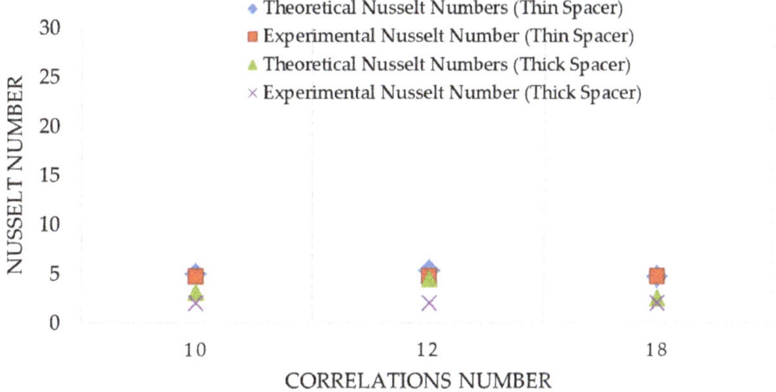

Figure 17. Laminar flow correlation for spacer-filled channels with flow rate (3 L/min).

As discussed above, most of the heat transfer correlations for spacer-filled channels that are recommended for localized turbulence (eddy flow) do not have a good fit with the experimental results from the DCMD module used for this work. Hence, it was decided to estimate the bulk flow conditions based on bulk Reynolds number. Based on the bulk Reynolds number for the experimental results used in this research, the flow regime is identified as laminar, so it was decided to compare the experimental results of the spacer filled channel with the correlations for empty channels (Table 1). For the 1.5 mm spacer, the most appropriate Nusselt number value was achieved using Equations (10) and (12), with a deviation ranging between 5% and 10%. However, considering 1.5 and 2 mm spacers, the heat transfer correlation for laminar flow given by Equation (18) resulted in the smallest deviation between theoretical and experimental results, at less than 15% for all spacer geometries and flow rates tested, as can be seen in Figure 17.

5. Conclusions

The heat transfer within a DCMD module separated by a thin plastic sheet instead of a hydrophobic membrane was analyzed to investigate the applicability of theoretical heat transfer correlations in two cases: empty channels and spacer-filled channels. For empty channels with flow rates of 3 and 5 L/min, the best correlation was provided by Equation (12). The results obtained using this correlation showed a deviation of less than 10% from the experimental results.

The spacer-filled channel cases investigation included two spacer thicknesses (1.5 and 2 mm) and flow rates of 3 and 5 L/min. Other heat transfer correlations for spacer-filled channels were applied, but the lowest deviation between experimental and theoretical results for these correlations was 42%. The experimental results for spacer-filled channels were also compared with the correlations for empty channels. It was found that the correlation given by Equation (18) had the lowest deviation, of less than 15%, for all spacer geometries and flow rates. As a result, the primary benefit of the outcomes of this study comes from identification of the best Nusselt number correlation for use in modelling heat transfer in DCMD systems, assuming similar channel geometries, flow rates and spacer geometries.

The other major outcome based on the experimental results is that including a mesh spacer material in the feed and permeate channels significantly increases the aspect ratio of the channel encouraging eddy flow conditions. This helps with the reduction in the temperature polarization. It should be noted that both 1.5 mm and 2 mm thick spacers resulted in similar improvements (for a 2.8 mm thick channel), with the thicker spacer giving slightly higher heat transfer. However, given the looseness of the fit within the channel for the 1.5 mm spacer, it could not be determined whether these improvements were the result of the same or different underlying mechanisms, so this is recommended as an area for future research.

Author Contributions: Conceptualization, H.F., A.A. and A.D.; data curation, H.F.; formal analysis, H.F. and R.K.; investigation, H.F.; methodology, H.F., M.B. and Q.L.V.; writing—original draft, H.F. and A.D.; supervision, A.A. and A.D., writing—review and editing, H.F., M.B., Q.L.V., O.T., R.K., A.A. and A.D. All authors have read and agreed to the published version of the manuscript.

Funding: This research received no external funding.

Data Availability Statement: The data is an experimental outcome and has been stored with the first author. Data will be made available on request by sending email to hzfaqeha@uqu.edu.sa.

Acknowledgments: Contributors are the coauthors of the manuscript.

Conflicts of Interest: The authors declare no conflict of interest.

Nomenclature

The following abbreviations are used in this manuscript:

A_m	The membrane's active surface area (m^2)
C_p	Specific heat capacity (J/kg·K)
D	Hydraulic diameter (m)
d_f	Filament size (m)
d_{hs}	Hydraulic diameter or the spacer-filled (m)
H	Spacer thickness (m)
h_f	Heat transfer coefficient at feed side (W/m^2·°C).
h_p	Heat transfer coefficient at permeate side (W/m^2·°C)
h_{fg}	Enthalpy of vaporization for water (J/kg)
k	Thermal conductivity (W/m^2·°C).
k_{dc}	Spacer geometry correction factor
l_m	Mesh size of spacer (m)
$\dot{m}$	Flow rate (kg/s)
Nu	Nusselt number
P_r	Prandtl number.
$\dot{Q}_f$	Heat transfer by convection at feed side (W)
$\dot{Q}_p$	Heat transfer by convection at permeate side (W)
$\dot{Q}_{cond}$	Heat transfer by conduction through the membrane (W)
R_e	Reynolds number.
T_f	Average bulk temperature of the feed stream (°C)
T_p	Average bulk temperature of the permeate stream (°C)

$T_{sheet,f}$	Feed side sheet surface temperature (°C)
$T_{sheet,p}$	Permeate side sheet surface temperature (°C)
ΔT_{LMTD}	Log mean temperature difference (°C)
U	The overall heat transfer coefficient (W/m^2·°C)
u	velocity (m/s)
ρ	Water density (kg/m^3)
δ	Membrane thickness (m)
θ	Hydrodynamic angle
ε	Spacer void
l_m	Mesh size of spacer (m)

References

1. Irannezhad, M.; Ahmadi, B.; Liu, J.; Chen, D.; Matthews, J.H. Global water security: A shining star in the dark sky of achieving the sustainable development goals. *Sustain. Horiz.* **2021**, *1*, 100005. [CrossRef]
2. Davis, A.C.; Arnocky, S.; Stroink, M. The problem of overpopulation: Proenvironmental concerns and behavior predict reproductive attitudes. *Ecopsychology* **2019**, *11*, 92–100. [CrossRef]
3. Membrane Distillation. Available online: https://emis.vito.be/en/bat/tools-overview/sheets/membrane-distillation (accessed on 15 October 2022).
4. El-Bourawi, M.S.; Ding, Z.; Ma, R.; Khayet, M. A framework for better understanding membrane distillation separation process. *J. Membr. Sci.* **2006**, *285*, 4–29. [CrossRef]
5. Alves, V.D.; Coelhoso, I. Orange juice concentration by osmotic evaporation and membrane distillation: A comparative study. *J. Food Eng.* **2006**, *74*, 125–133. [CrossRef]
6. Calabro, V.; Jiao, B.L.; Drioli, E. Theoretical and Experimental Study on Membrane Distillation in the Concentration of Orange Juice. *Ind. Eng. Chem. Res.* **1994**, *33*, 1803–1808. [CrossRef]
7. Martínez-Díez, L.; Vázquez-González, M.; Florido-Díaz, F. Study of membrane distillation using channel spacers. *J. Membr. Sci.* **1998**, *144*, 45–56. [CrossRef]
8. Taamneh, Y.; Bataineh, K. Improving the performance of direct contact membrane distillation utilizing spacer-filled channel. *Desalination* **2017**, *408*, 25–35. [CrossRef]
9. Phattaranawik, J.; Jiraratananon, R.; Fane, A.; Halim, C. Mass flux enhancement using spacer filled channels in direct contact membrane distillation. *J. Membr. Sci.* **2001**, *187*, 193–201. [CrossRef]
10. Phattaranawik, J.; Jiraratananon, R.; Fane, A. Effects of net-type spacers on heat and mass transfer in direct contact membrane distillation and comparison with ultrafiltration studies. *J. Membr. Sci.* **2003**, *217*, 193–206. [CrossRef]
11. Yun, Y.; Wang, J.; Ma, R.; Fane, A.G. Effects of channel spacers on direct contact membrane distillation. *Desalination Water Treat.* **2011**, *34*, 63–69. [CrossRef]
12. Taamneh, Y. Heat and fluid flow through a spacer-filled channel in a membrane module. *Heat Transf. Res.* **2017**, *48*, 1567–1580. [CrossRef]
13. Chang, H.; Hsu, J.-A.; Chang, C.-L.; Ho, C.-D. CFD Study of Heat Transfer Enhanced Membrane Distillation Using Spacer-Filled Channels. *Energy Procedia* **2015**, *75*, 3213–3219. [CrossRef]
14. Gong, B.; Yang, H.; Wu, S.; Xiong, G.; Yan, J.; Cen, K.; Bo, Z.; Ostrikov, K. Graphene Array-Based Anti-fouling Solar Vapour Gap Membrane Distillation with High Energy Efficiency. *Nano-Micro Lett.* **2019**, *11*, 1–14. [CrossRef] [PubMed]
15. Ve, L.Q.; Rahaoui, K.; Bawahab, M.; Faqeha, H.; Date, A.; Akbarzadeh, A.; Cuong, D.M.; Nguyen, L.Q. Experimental Investigation of Heat Transfer Correlation for Direct Contact Membrane Distillation in Laminar Flow. *J. Heat Transf.* **2020**, *142*, 12001. [CrossRef]
16. Schofield, R.; Fane, A.; Fell, C. Heat and mass transfer in membrane distillation. *J. Membr. Sci.* **1987**, *33*, 299–313. [CrossRef]
17. Laganà, F.; Barbieri, G.; Drioli, E. Direct contact membrane distillation: Modelling and concentration experiments. *J. Membr. Sci.* **2000**, *166*, 1–11. [CrossRef]
18. Qtaishat, M.; Matsuura, T.; Kruczek, B.; Khayet, M. Heat and mass transfer analysis in direct contact membrane distillation. *Desalination* **2008**, *219*, 272–292. [CrossRef]
19. Franken, A.; Nolten, J.; Mulder, M.; Bargeman, D.; Smolders, C. Wetting criteria for the applicability of membrane distillation. *J. Membr. Sci.* **1987**, *33*, 315–328. [CrossRef]
20. Tijing, L.D.; Choi, J.S.; Lee, S.; Kim, S.H.; Shon, H.K. Recent progress of membrane distillation using electrospun nanofibrous membrane. *J. Membr. Sci.* **2014**, *453*, 435–462. [CrossRef]
21. Gryta, M.; Tomaszewska, M.J.J. Heat transport in the membrane distillation process. *J. Membr. Sci.* **1998**, *144*, 211–222. [CrossRef]
22. Sieder, E.N.; Tate, G.E. Heat Transfer and Pressure Drop of Liquids in Tubes. *Ind. Eng. Chem.* **1936**, *28*, 1429–1435. [CrossRef]
23. Thomas, L. *Heat Transfer*; Prentice-Hall, Inc.: Englewood Cliffs, NJ, USA, 1992.
24. Gryta, M.; Tomaszewska, M.; Morawski, A.W. Membrane distillation with laminar flow. *Sep. Purif. Technol.* **1997**, *11*, 93–101. [CrossRef]
25. Andrjesdóttir, O.; Ong, C.L.; Nabavi, M.; Paredes, S.; Khalil, A.; Michel, B.; Poulikakos, D. An experimentally optimized model for heat and mass transfer in direct contact membrane distillation. *Int. J. Heat Mass Transf.* **2013**, *66*, 855–867. [CrossRef]

26. Chang, H.; Ho, C.-D.; Hsu, J.-A. Analysis of Heat Transfer Coefficients in Direct Contact Membrane Distillation Modules Using CFD Simulation. *J. Appl. Sci. Eng.* **2016**, *19*, 197–206. [CrossRef]
27. Da Costa, A.; Fane, A.; Wiley, D. Spacer characterization and pressure drop modelling in spacer-filled channels for ultrafiltration. *J. Membr. Sci.* **1994**, *87*, 79–98. [CrossRef]
28. Da Costa, A.R. Fluid Flow and Mass Transfer in Spacer-Filled Channels for Ultrafiltration. Ph.D. Thesis, University of New South Wales, Sydney, Australia, 1993.
29. Schock, G.; Miquel, A. Mass transfer and pressure loss in spiral wound modules. *Desalination* **1987**, *64*, 339–352. [CrossRef]
30. Schwager, F.; Robertson, P.; Ibl, N. The use of eddy promoters for the enhancement of mass transport in electrolytic cells. *Electrochim. Acta* **1980**, *25*, 1655–1665. [CrossRef]
31. Zhang, J.; Gray, S.; Li, J.-D. Modelling heat and mass transfers in DCMD using compressible membranes. *J. Membr. Sci.* **2012**, *387–388*, 7–16. [CrossRef]
32. Heaton, C.J. On the appearance of Moffatt eddies in viscous cavity flow as the aspect ratio varies. *Phys. Fluids* **2008**, *20*, 103102. [CrossRef]
33. Phattaranawik, J.; Jiraratananon, R.; Fane, A.G. Heat transport and membrane distillation coefficients in direct contact membrane distillation. *J. Membr. Sci.* **2003**, *212*, 177–193. [CrossRef]
34. Kim, Y.-D.; Francis, L.; Lee, J.-G.; Ham, M.-G.; Ghaffour, N. Effect of non-woven net spacer on a direct contact membrane distillation performance: Experimental and theoretical studies. *J. Membr. Sci.* **2018**, *564*, 193–203. [CrossRef]

Article

Thermal Performance of Integrated Direct Contact and Vacuum Membrane Distillation Units

Alessandra Criscuoli

Institute on Membrane Technology (CNR-ITM), via P. Bucci 17/C, 87036 Rende, CS, Italy; a.criscuoli@itm.cnr.it; Tel.: +39-0984-492118

Abstract: An integrated membrane distillation (MD) flowsheet, consisting of direct contact membrane distillation (DCMD) and vacuum membrane distillation (VMD) units, was proposed and analysed in terms of thermal performance and water recovery factor, for the first time. The same lab-scale membrane module (40 cm^2) was used for carrying out experiments of DCMD and VMD at fixed feed operating conditions (deionised water at 230 L/h and ~40 °C) while working at the permeate side with deionised water at 18 °C and with a vacuum of 20 mbar for the DCMD and the VMD configuration, respectively. Based on experimental data obtained on the single modules, calculations of the permeate production, the specific thermal energy consumption (STEC) and the gained output ratio (GOR) were carried out for both single and integrated units. Moreover, the calculations were also made for a flow sheet consisting of two DCMD units in series, representing the "traditional" way in which more units of the same MD configuration are combined to enhance the water recovery factor. A significant improvement of the thermal performance (lower STEC and higher GOR) was obtained with the integrated DCMD–VMD flowsheet with respect to the DCMD units operating in series. The integration of DCMD with VMD also led to a higher permeate production and productivity/size (PS) ratio, a metric defined to compare plants in terms of the process intensification strategy.

Keywords: thermal performance; water recovery factor; direct contact membrane distillation (DCMD); vacuum membrane distillation (VMD); integrated MD units

Citation: Criscuoli, A. Thermal Performance of Integrated Direct Contact and Vacuum Membrane Distillation Units. *Energies* **2021**, *14*, 7405. https://doi.org/10.3390/en14217405

Academic Editors: Sanghyun Jeong and Albert S. Kim

Received: 1 September 2021
Accepted: 4 November 2021
Published: 6 November 2021

Publisher's Note: MDPI stays neutral with regard to jurisdictional claims in published maps and institutional affiliations.

1. Introduction

The potential of membrane distillation has been successfully investigated in different fields of industrial interest, such as wastewater treatment, desalination, agro-food and beverage, and biomedical applications [1–7]. MD is able to well reject all non volatile compounds contained into the water stream to be treated, thanks to the fact that water is not removed through the membrane as liquid, but as vapor. More specifically, the aqueous feed is in contact at atmospheric pressure with one side of a microporous hydrophobic membrane and it is not allowed to pass through the membrane. Then, to promote the liquid evaporation, the feed is usually heated (typical temperatures range from 40 °C to 80 °C) while at the other side of the membrane:

(i)   A colder aqueous stream (permeate stream) is sent at atmospheric pressure, which is blocked at the membrane hydrophobic surface (DCMD configuration);
(ii) An air gap is created between the membrane and a condensing surface (air gap membrane distillation: AGMD configuration);
(iii) A vacuum is applied (VMD configuration);
(iv) A cold sweep gas is sent (sweep gas membrane distillation: SGMD configuration).

A common drawback of all MD configurations is the need for thermal energy to heat the feed and to keep it at the desired temperature throughout the process. One of the main limitations of the industrial implementation of MD is, in fact, linked to its high thermal demand [8–10]. In this respect, in addition to the use of renewable energies as heat sources for the MD unit [11–13], the development of modules with internal heat recovery [14–16] is in progress. More recently, membranes with localized heating [17–21] have also been investigated. Among the MD configurations previously reported, which are those most-assessed, DCMD is the simplest and the most investigated one. In DCMD, during the evaporation the feed becomes colder while the permeate increases its temperature because of the vapor condensation. With this configuration, the heat recovery is not possible inside the module, and it is therefore performed in an external heat exchanger where the feed is pre-heated by the permeate stream, which, in turns, is pre-cooled. The final heat of the feed and the cool of the permeate occur in two other external heat exchangers, so as to recirculate both streams at the desired operating temperatures [22–24]. It has to be pointed out that both the internal and the external heat recovery systems need feed temperatures higher than 50 °C to operate with reasonable efficiency and, therefore, MD applications at lower temperatures are not well covered. On the other hand, membranes with localized heating, which can also be applied at low operating temperatures, are still at the development stage. Considering the relevant number of studies on DCMD, this paper focuses on a possible alternative strategy to improve the thermal performance of DCMD when operating at low feed temperatures (~40 °C). It is known that in MD the water recovery factor per pass is quite low (maximum 8% [25]) and, therefore, membrane distillation plants must work with feed recirculation and with different modules in series, in order to ensure an acceptable productivity. Usually, the same membrane distillation configuration is considered for the modules which work in series.

In this work, it was proposed to couple two different MD configurations, so as to improve not only the water recovery factor, but also the thermal performance. In particular, for the first time to the best of the author's knowledge, it was proposed to couple the DCMD unit with a VMD one, so that the DCMD retentate is processed as feed in VMD before being recirculated back to the DCMD module. VMD was chosen because it is able to produce high trans-membrane fluxes also at low feed temperatures and it avoids the heat loss by conduction through the membrane material.

DCMD and VMD have been compared in literature for different applications. Table 1 gives an overview of the main studies carried out.

Table 1. Comparison between DCMD and VMD reported in the literature.

Membrane	Feed	J (kg/m²h)		Reference
		DCMD	**VMD**	
PVDF-HF	Salty solution	~16	~24	[26]
PVDF-FS	Salty solution	~32	~40	[27]
PVDF-HF	Distilled water	~8	~42	[28]
PP-HF	Distilled water	~5	~22	[28]
Si$_3$N$_4$-HF	Salty solution	~13.5	~36	[29]
PP-HF	OMWW	~6.5	~19	[30]
PTFE-FS	Dyes solution	~17.4	~37.4	[31]
Ceramic-T	Salty solution	~4	~25	[32]
PP-HF	Carbonate solution	~0.11	~0.8	[33]

PTFE: polytetrafluoroethylene; HF: hollow fibre; FS: flat sheet; T: tubular; OMWW: olive mill wastewater.

In particular, the reported flux values are those obtained by working in the two MD configurations at the same feed temperature and concentration. Membranes used were mainly in polyvinylidene fluoride (PVDF) and polypropylene (PP). However, due to the difference in membrane and module features, as well as the difference in the operating conditions and in the treated feeds, trans-membrane fluxes varied for the different studies. Nevertheless, in all cases, the VMD configuration led to a higher flux than the DCMD one.

The higher efficiency in permeate production of VMD was also confirmed in a theoretical work of Guan et al. [34] who simulated the performance of DCMD and VMD to treat a 7 wt% NaCl feed in a hollow fibre membrane distillation module. By working at the same operating conditions, VMD led to a 2.5-fold increase in the permeate stream and to a reduction of the specific energy consumption.

In this work, the proposed integrated DCMD–VMD flowsheet was compared in terms of STEC, GOR and permeate production, with a flowsheet consisting of two DCMD units in series, representing the "traditional" way in which more units of the same MD configuration are combined to enhance the water recovery factor. The analysis did not include the electrical energy consumptions, which are known to contribute to the overall energy consumption of MD much less than the thermal ones. For instance, in the work of Méricq et al. [35], the heat energy demand for VMD applied to seawater desalination is more than 98% of the total energy requirements.

2. Materials and Methods

2.1. Experimental Lab Set-Ups

The DCMD and VMD tests were carried out on the same lab module (40 cm^2 membrane area) where a flat commercial polypropylene membrane (0.2 μm pore size, 70% porosity, 91 μm thickness) purchased from Membrana (Germany), now 3M, was used. All tests were carried out on deionised water as feed, to make a more general analysis, then, avoiding any potential issues linked to a specific feed to be treated. Deionised water was sent to the bottom plate of the module in both cases and the feed-side was operated under the same conditions of flow rate and temperature. In particular, the flowrate was set at the highest value reachable by the pump (230 L/h, corresponding to a velocity of 0.46 m/s inside the module), in order to reduce the heat transfer resistance of the boundary layer, to promote more mixing, so as reducing fouling issues in applications with real aqueous streams, and to decrease the feed temperature decay along the module, by reducing its residence time. The feed temperatures at the inlet and outlet of the module were read by thermocouples. At the top plate, a cold deionised stream was sent in counter-current during DCMD experiments and the permeated vapor was directly condensed at the membrane-cold interface, while a vacuum was applied for VMD runs. In the latter configuration, the top plate had only one exit active for the permeate removal and its condensation occurred in a trap immersed into liquid nitrogen. Liquid nitrogen was used to ensure that all water vapor was condensed before the vacuum pump. It has to be noticed that this step can also be made using cooling fluids in specifically designed condensers. Figures 1 and 2 show the schemes of the DCMD and VMD lab set-ups, respectively.

In order to operate at a similar driving force during the VMD and DCMD experiments, a vacuum of 20 mbar was applied at the permeate side in VMD, while in DCMD the cold stream was recirculated at 18 °C temperature, at which the water vapor pressure is 19.7 mbar. The main operating conditions used are summarized in Table 2.

Table 2. Main operating conditions used during DCMD and VMD tests.

	DCMD	VMD
Q_f (L/h)	230	230
T_f (°C)	39.6	39.6
Q_d (L/h)	200	/
T_d (°C)	18	/
P_v (mbar)	/	20

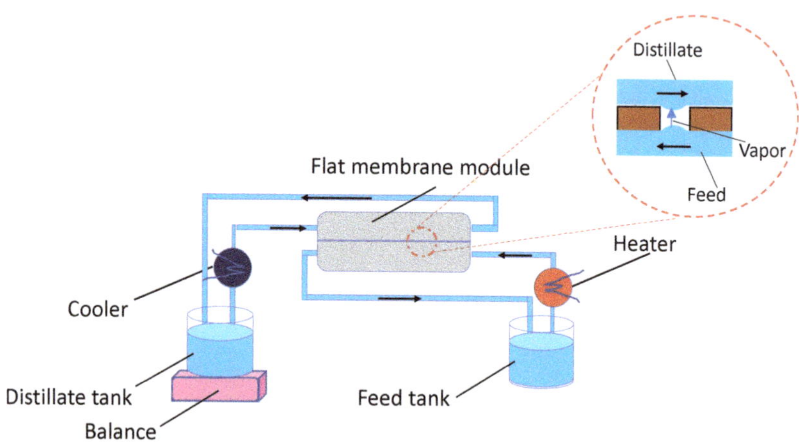

Figure 1. Scheme of the DCMD lab set-up and sketch of the vapor transport through a micropore.

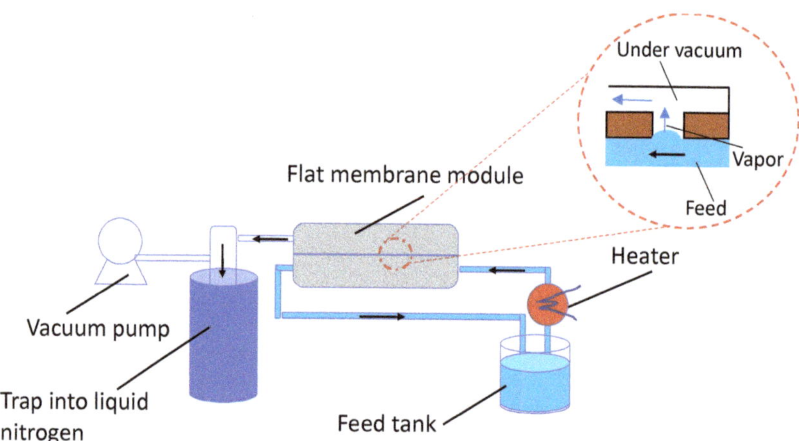

Figure 2. Scheme of the VMD lab set-up and sketch of the vapor transport through a micropore.

2.2. Followed Methodology

Based on the tests carried out on the single DCMD and VMD units, the related trans-membrane flux and decay of feed temperature along the module for the two MD configurations were obtained. Then, the experimental values were used to analyse each unit in terms of permeate production, thermal energy consumption and efficiency of evaporation (see the following sections for details on their calculation). The same analysis was then extended to the DCMD–DCMD and integrated DCMD–VMD flowsheets, which were proposed considering the experimental results. Finally, the integrated DCMD–VMD flowsheet was compared, also in terms of size, with the two DCMD units in series. Figure 3 summarizes the followed steps.

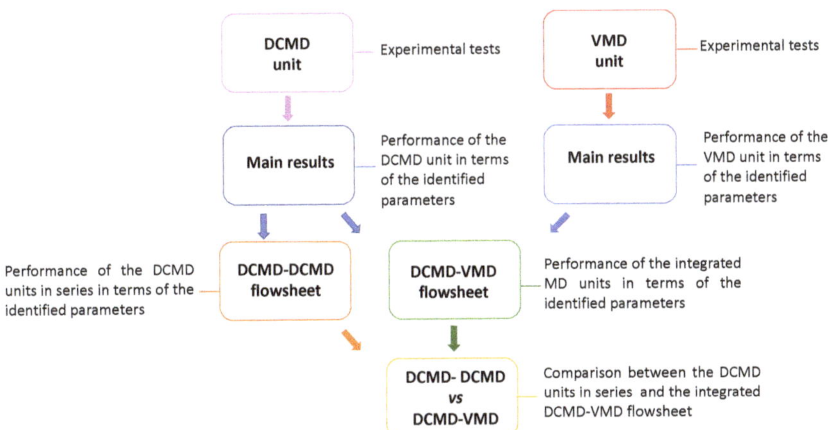

Figure 3. Summary of the steps followed in this work.

2.3. Flux and Permeate Flow Rate Calculation

In DCMD, the mass of the permeate was weighed by a balance located under the distillate tank. In VMD, the permeate was first recovered as liquid into the trap and afterwards weighed. In both cases, the flux (J) was calculated by dividing the permeate mass (m, g) by the membrane area (A_m, m^2) and the experimental time (t, h):

$$J = \frac{m}{A_m \times t} \tag{1}$$

The permeate flow rate (Q_p, g/h) was obtained by multiplying the permeate flux by the membrane area:

$$Q_p = J \times A_m \tag{2}$$

2.4. Specific Thermal Energy Consumption and Gained Output Ratio Calculation

The STEC represents the thermal energy consumption (Q_H) associated with a certain permeate production and it is given by the ratio between the total thermal energy consumption (J/h) and the permeate flow rate:

$$STEC = \frac{Q_H}{Q_p} \tag{3}$$

The GOR gives an indication of how much of the thermal energy consumption (Q_H) is used to produce the permeate (evaporation efficiency), and it can be calculated by the ratio between the thermal energy effectively used for the evaporation and the total thermal energy consumption:

$$GOR = \frac{Q_p \times \Delta H_v}{Q_H} \tag{4}$$

In the above formula, ΔH_v is the enthalpy of vaporization (J/g).
Q_H is given by the sum of the single thermal energies (Q_{Hi}) supplied to the plant:

$$Q_H = \sum_{i=1}^{n} Q_{Hi} \tag{5}$$

Each thermal energy is calculated as function of the properties of the stream i, such as its flow rate Q_{fi} (g/h), its specific heat capacity cp_i (J/gK) and its difference of temperature in the plant ΔT_i (K):

$$Q_{Hi} = Q_{fi} \times cp_i \times \Delta T_i \tag{6}$$

2.5. Productivity/Size Ratio Calculation

The PS ratio is a metric introduced to compare plants in terms of the process intensification strategy parameters. In this strategy, an important aspect of future plants is to work with higher productivity and reduced size [36]. In this respect, the PS ratio aims to compare the ratio of the productivity and size of two plants [37] and in this study was used to compare the investigated membrane distillation flowsheets by means of the following formula:

$$PS = \frac{\text{Productivity}/\text{Size}\Big| MD_{\text{flowsheet integrated units}}}{\text{Productivity}/\text{Size}\Big| MD_{\text{flowsheet DCMD units in series}}} \tag{7}$$

If PS is higher than 1, the MD flowsheet where the different MD units are integrated (DCMD–VMD) must be preferred.

3. Results and Discussion

3.1. Single DCMD and VMD Units

Figure 4 shows the trans-membrane fluxes measured and the variations of the feed temperature ($\Delta Tf = T_{fin} - Tf_{out}$) registered for the two MD units. The experiments were repeated to ensure the reproducibility of the results and the average flux values are reported. A significant difference in flux was observed, even though the same driving force was applied. Being that the fluid dynamic at the feed-side is the same for the two MD configurations, the difference in flux can be attributed to an increase in the vapor pressure at the membrane-cold side of the DCMD unit. In fact, the distillate temperature at the membrane surface can be higher than that of the bulk, due to temperature polarization phenomena, which did not allow a fast removal of both the condensation heat and the heat transferred by conduction through the membrane polymer from the feed to the distillate side [28]. Moreover, due to the negligible conductive heat loss in VMD, a higher driving force is established during the VMD tests [34]. Finally, the vapor transport through the micropores is usually regulated by a combination of the Knudsen and molecular diffusion mechanisms in DCMD, whilst the Knudsen mechanism dominates in VMD [38]. Concerning the decay of the feed temperature along the module, a slightly higher value was registered for the DCMD tests, probably because of the heat lost by conduction through the polymer.

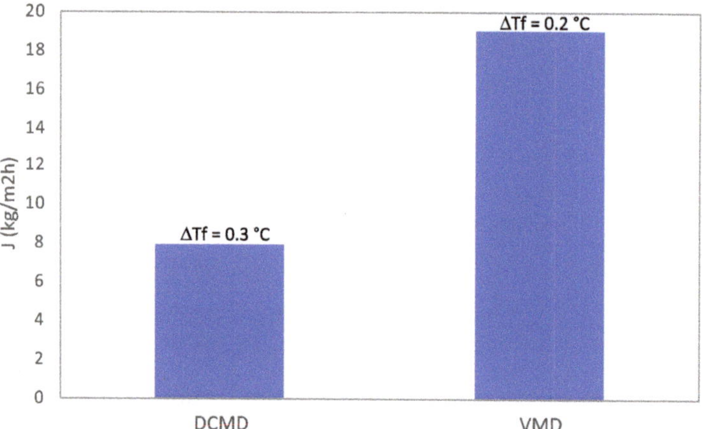

Figure 4. Trans-membrane flux and feed temperature variation for the two MD units.

The obtained results are in agreement with the literature data reported in Table 1, confirming the higher productivity of the VMD configuration at parity of feed-side conditions.

Based on the operating conditions and the obtained data, the flowsheets for the two MD configurations were sketched, as reported in Figures 5 and 6. In particular, each flowsheet includes the flow rates and the temperatures of the feed streams, together with the thermal supplies needed. From them, the main parameters for comparing the DCMD and the VMD performance were calculated by Equations (2)–(4), and they are listed in Table 3. It is evident that the VMD unit was more efficient, leading to a higher permeate production and GOR and to a lower STEC. The main reason for these positive results lies in the much higher permeate flux achievable with the VMD configuration.

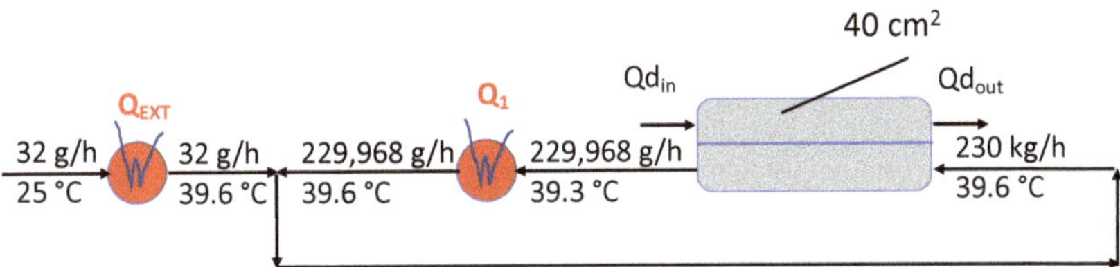

Figure 5. DCMD flowsheet.

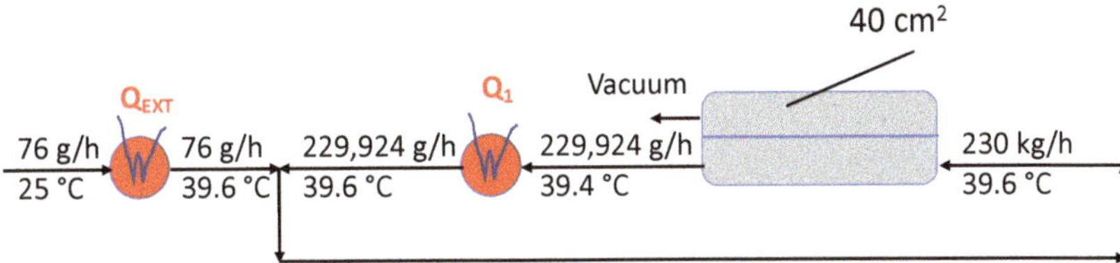

Figure 6. VMD flowsheet.

Table 3. DCMD and VMD performances.

Configuration	Qp (g/h)	STEC (W/g/h)	GOR
DCMD	32	2.52	0.27
VMD	76	0.72	0.93

In fact, by comparing the STEC and GOR of the two membrane operations at parity of feed temperature decay along the module, the VMD unit always led to a better performance, as reported in Figures 7 and 8, thanks to the more than double permeate production with respect to the DCMD configuration. In particular, when the same temperature decay registered in the VMD tests was considered (0.2 °C), the STEC of the DCMD unit decreased from 2.52 W/g/h to 1.69 W/g/h, while the GOR increased from 0.27 to 0.4. When the feed temperature decay measured in the DCMD tests was applied (0.3 °C), the STEC and the GOR of the VMD unit moved from 0.72 W/g/h to 1.07 W/g/h and from 0.93 to 0.62, respectively.

3.2. DCMD Units in Series

Based on the experimental results on the DCMD unit, a flowsheet where two DCMD units work in series was analysed, as reported in Figure 9. Both DCMD units worked in the same manner as the experimental one, since the stream exiting the first DCMD unit was heated up to the experimental temperature (39.6 °C) before entering the second DCMD unit. The calculated STEC and GOR of the flowsheet were the same as those of the single DCMD unit, the only difference being the higher membrane area (80 cm^2) and permeate production (64 g/h).

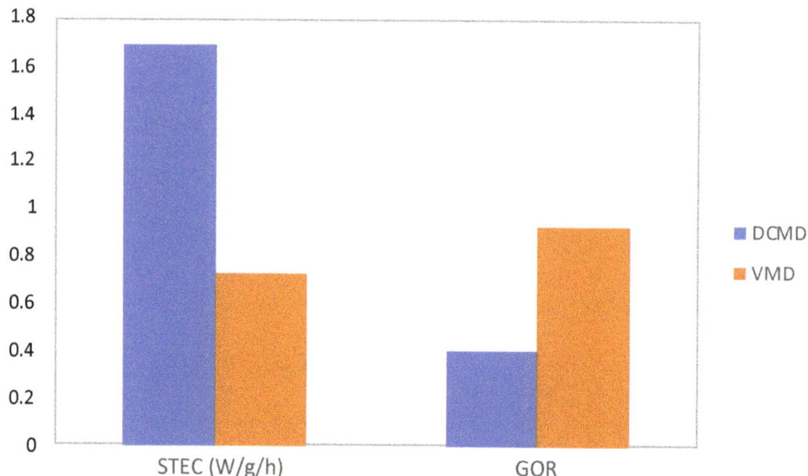

Figure 7. STEC and GOR of the two MD units, considering for both the experimental feed temperature decay of the VMD unit (0.2 °C).

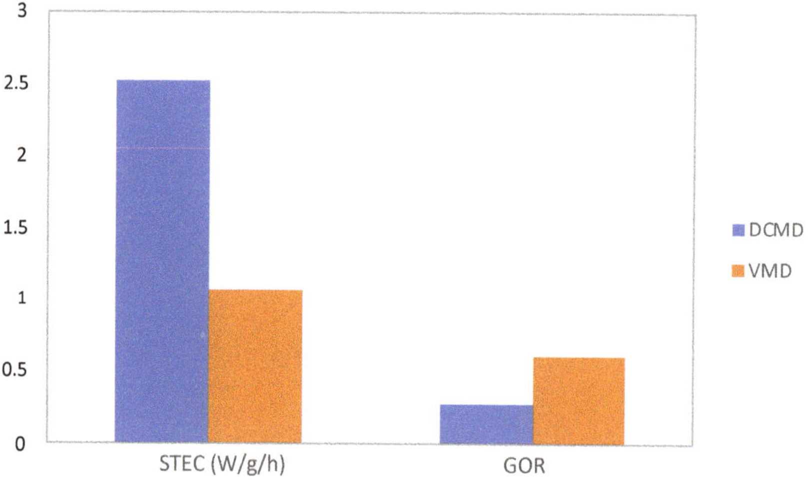

Figure 8. STEC and GOR of the two MD units, considering for both the experimental feed temperature decay of the DCMD unit (0.3 °C).

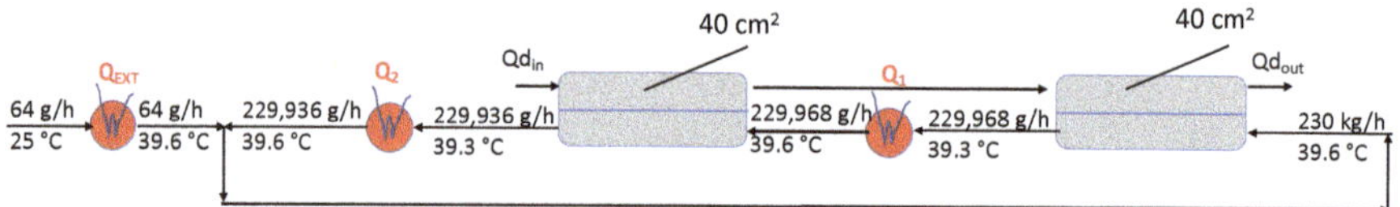

Figure 9. DCMD–DCMD flowsheet.

3.3. Integrated DCMD–VMD Units

After having analysed the single MD units, an integrated DCMD–VMD flowsheet was investigated, with the aim of boosting the DCMD thermal performance, taking advantage of the higher efficiency of VMD. In this case, the feed stream exiting the DCMD unit was sent to the VMD one before being recycled back to the DCMD module (see Figure 10).

As for the DCMD units in series, in the integrated configuration, both DCMD and VMD units operated under the same experimental conditions used in the single DCMD and VMD flowsheets, respectively.

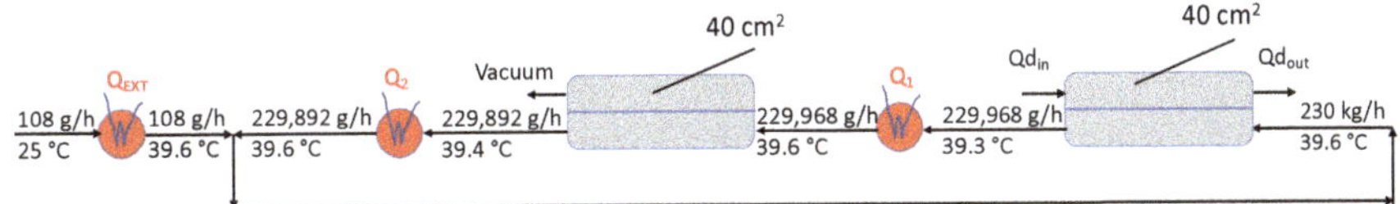

Figure 10. Integrated DCMD–VMD flowsheet.

Table 4 compares the flowsheet with the DCMD units in series to the DCMD–VMD one in terms of permeate production. At parity of the membrane area, the integrated flowsheet led to a substantial increment of productivity, moving from 64 g/L to 108 g/L. It also showed lower STEC and higher GOR values (see Figure 11), thus improving the thermal performance of the DCMD unit.

Table 4. Comparison between the flowsheet with the DCMD units in series and the integrated DCMD–VMD flowsheet in terms of permeate production.

Configuration	Qp (g/h)
DCMD–DCMD	64
DCMD–VMD	108

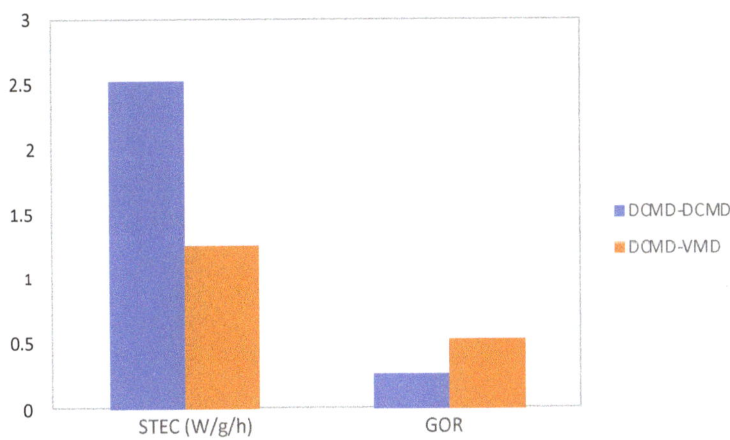

Figure 11. Comparison between the flowsheet with the DCMD units in series and the integrated DCMD–VMD flowsheet in terms of STEC and GOR.

Specifically, gains obtained with the integrated DCMD–VMD flowsheet are summarized in Table 5. It is clear that, despite the same membrane area, the Productivity/Size ratio is in favour of the new proposed integrated MD configuration, thanks to the significant increase in the permeate production.

Table 5. Gain obtained with the integrated DCMD–VMD flowsheet.

Q_P Increase (%)	STEC Reduction (%)	GOR Increase (%)	PS Ratio (/)
69	50	100	1.69

4. Conclusions

For the first time, as an alternative strategy to improve the thermal performance and the water recovery factor of DCMD, the integration of DCMD and VMD was proposed and investigated. In particular, the analysis was carried out at low operating temperatures ($\sim$40 °C), for which the existing heat recovery methods are not very efficient, since they usually work at temperatures higher than 50 °C. Calculations were also made for a flow sheet consisting of two DCMD units in series, representing the "traditional" way in which more units of the same MD configuration are combined to enhance the water recovery factor.

The integrated DCMD–VMD flowsheet led to significant benefits with respect to the DCMD–DCMD units:

(i)   Reduction of the STEC (by 50%);
(ii)   Increase in the GOR (by 100%);
(iii)   Increase in the permeate production (by 69%);
(iv)   Higher productivity per membrane area (PS = 1.69).

On the basis of the above considerations, the application of a VMD unit in series with the DCMD one resulted to be an interesting way to improve both thermal performance and plant productivity, while fitting well the requirements of the process intensification strategy. It has to be pointed out that, although this work was carried out to cover the application of DCMD for treating feeds that need low operating temperatures, the same concept can be extended to higher operating temperatures and, then, to a wider number of processes. In this case, a comparison of the thermal performance of the proposed integrated MD flowsheet with that of existing heat recovery methods must be made, in order to identify the most effective one.

Funding: This research received no external funding.

Conflicts of Interest: The author declares no conflict of interest.

References

1.   Chiam, C.-K.; Sarbatly, R. Vacuum membrane distillation processes for aqueous solution treatment—A review. *Chem. Eng. Process. Process. Intensif.* **2013**, *74*, 27–54. [CrossRef]
2.   Abu-Zeid, M.A.E.-R.; Zhang, Y.; Dong, H.; Zhang, L.; Chen, H.-L.; Hou, L. A comprehensive review of vacuum membrane distillation technique. *Desalination* **2015**, *356*, 1–14. [CrossRef]
3.   Wang, P.; Chung, T.-S. Recent advances in membrane distillation processes: Membrane development, configuration design and application exploring. *J. Membr. Sci.* **2015**, *474*, 39–56. [CrossRef]
4.   Ashoor, B.B.; Mansour, S.; Giwa, A.; Dufour, V.; Hasan, S.W. Principles and applications of direct contact membrane distillation (DCMD): A comprehensive review. *Desalination* **2016**, *398*, 222–246. [CrossRef]
5.   Tavakkoli, S.; Lokare, O.; Vidic, R.; Khanna, V. A techno-economic assessment of membrane distillation for treatment of Marcellus shale produced water. *Desalination* **2017**, *416*, 24–34. [CrossRef]
6.   Chen, Y.R.; Chen, L.H.; Chen, C.H.; Ko, C.C.; Huang, A.; Li, C.L.; Chuang, C.J.; Tung, K.L. Hydrophobic alumina hollow fiber membranes for sucrose concentration by vacuum membrane distillation. *J. Membr. Sci.* **2018**, *555*, 250–257. [CrossRef]
7.   Khumalo, N.; Nthunya, L.; Derese, S.; Motsa, M.; Verliefde, A.; Kuvarega, A.; Mamba, B.B.; Mhlanga, S.; Dlamini, D.S. Water recovery from hydrolysed human urine samples via direct contact membrane distillation using PVDF/PTFE membrane. *Sep. Purif. Technol.* **2019**, *211*, 610–617. [CrossRef]
8.   Criscuoli, A. Improvement of the membrane distillation performance through the integration of different configurations. *Chem. Eng. Res. Des.* **2016**, *111*, 316–322. [CrossRef]
9.   Deshmukh, A.; Boo, C.; Karanikola, V.; Lin, S.; Straub, A.P.; Tong, T.; Warsinger, D.M.; Elimelech, M. Membrane distillation at the water-energy nexus: Limits, opportunities, and challenges. *Energy Environ. Sci.* **2018**, *11*, 1177–1196. [CrossRef]
10.   Lokare, O.R.; Tavakkoli, S.; Khanna, V.; Vidic, R.D. Importance of feed recirculation for the overall energy consumption in membrane distillation systems. *Desalination* **2018**, *428*, 250–254. [CrossRef]
11.   Raluy, G.R.; Schwantes, R.; Subiela, V.J.; Penate, B.; Melian, G.; Betancort, J. Operational experience of a solar membrane distillation demonstration plant in Pozo Izquierdo-Gran Canaria Island (Spain). *Desalination* **2012**, *290*, 1–13. [CrossRef]
12.   Saffarini, R.B.; Summers, E.K.; Arafat, H.A.; Leinhard, J.H.V. Technical evaluation of stand-alone solar powered membrane distillation systems. *Desalination* **2012**, *286*, 332–341. [CrossRef]
13.   Suárez, F.; Ruskowitz, J.A.; Tyler, S.W.; Childress, A.E. Renewable water: Direct contact membrane distillation coupled with solar ponds. *Appl. Energy* **2015**, *158*, 532–539. [CrossRef]

14. Winter, D.; Koschikowski, J.; Wieghaus, M. Desalination using membrane distillation: Experimental studies on full scale spiral wound modules. *J. Membr. Sci.* **2011**, *375*, 104–112. [CrossRef]

15. Zhao, K.; Heinzl, W.; Wenzel, M.; Buttner, S.; Bollen, F.; Lange, G.; Heinzl, S.; Sarda, N. Experimental study of the memsys vacuum-multi-effect-membrane-distillation (V-MEMD) module. *Desalination* **2013**, *323*, 150–160. [CrossRef]

16. Jansen, A.E.; Assink, J.W.; Hanemaaijer, J.H.; van Medevoort, J.; van Sonsbeek, E. Development and pilot testing of full-scale membrane distillation modules for deployment of waste heat. *Desalination* **2013**, *323*, 55–65. [CrossRef]

17. Wu, X.; Jiang, Q.; Ghim, D.; Singamaneni, S.; Jun, Y.-S. Localized heating with a photothermal polydopamine coating facilitates a novel membrane distillation process. *J. Mater. Chem. A* **2018**, *6*, 18799–18807. [CrossRef]

18. Huang, Q.; Gao, S.; Huang, Y.; Zhang, M.; Xiao, C. Study on photothermal PVDF/ATO nanofiber membrane and its membrane distillation performance. *J. Membr. Sci.* **2019**, *582*, 203–210. [CrossRef]

19. Alsaati, A.; Marconnet, A.M. Energy efficient membrane distillation through localized heating. *Desalination* **2018**, *442*, 99–107. [CrossRef]

20. Tan, Y.Z.; Han, L.; Chew, N.G.P.; Chow, W.H.; Wang, R.; Chew, J.W. Membrane distillation hybridized with a thermoelectric heat pump for energy-efficient water treatment and space cooling. *Appl. Energy* **2018**, *231*, 1079–1088. [CrossRef]

21. Ahmed, F.E.; Lalia, B.S.; Hashaikeh, R.; Hilal, N. Enhanced performance of direct contact membrane distillation via selected electrothermal heating of membrane surface. *J. Membr. Sci.* **2020**, *610*, 118224. [CrossRef]

22. Lee, H.Y.; He, F.; Song, L.M.; Gilron, J.; Sirkar, K.K. Desalination with a cascade of cross-flow hollow fiber membrane distillation devices integrated with a heat exchanger. *AICHE J.* **2011**, *57*, 1780–1795. [CrossRef]

23. Lin, S.; Yip, N.Y.; Elimelech, M. Direct contact membrane distillation with heat recovery: Thermodynamic insights from module scale modeling. *J. Membr. Sci.* **2014**, *453*, 498–515. [CrossRef]

24. Guan, G.; Yang, X.; Wang, R.; Fane, A.G. Evaluation of heat utilization in membrane distillation desalination system integrated with heat recovery. *Desalination* **2015**, *366*, 80–93. [CrossRef]

25. Swaminathan, J.; Lienhard, J.H.V. Design and operation of membrane distillation with feed recirculation for high recovery brine concentration. *Desalination* **2018**, *445*, 51–62. [CrossRef]

26. Tang, Y.; Li, N.; Liu, A.; Ding, S.; Yi, C.; Liu, H. Effect of spinning conditions on the structure and performance of hydrophobic PVDF hollow fiber membranes for membrane distillation. *Desalination* **2012**, *287*, 326–339. [CrossRef]

27. Fan, H.; Peng, Y. Application of PVDF membranes in desalination and comparison of the VMD and DCMD processes. *Chem. Eng. Sci.* **2012**, *79*, 94–102. [CrossRef]

28. Drioli, E.; Ali, A.; Simone, S.; Macedonio, F.; AL-Jlil, S.A.; Al Shabonah, F.S.; Al-Romaih, H.S.; Al-Harbi, O.; Figoli, A.; Criscuoli, A. Novel PVDF hollow fiber membranes for vacuum and direct contact membrane distillation applications. *Sep. Pur. Technol.* **2013**, *115*, 27–38. [CrossRef]

29. Zhang, J.W.; Fang, H.; Wang, J.W.; Hao, L.Y.; Xu, X.; Chen, C.S. Preparation and characterization of silicon nitride hollow fiber membranes for seawater desalination. *J. Membr. Sci.* **2014**, *450*, 197–206. [CrossRef]

30. Carnevale, M.C.; Gnisci, E.; Hilal, J.; Criscuoli, A. Direct Contact and Vacuum Membrane Distillation application for the olive mill wastewater treatment. *Sep. Purif. Technol.* **2016**, *169*, 121–127. [CrossRef]

31. Ramlow, H.; Machado, R.A.F.; Bierhalz, A.C.K.; Marangoni, C. Influence of dye class on the comparison of direct contact and vacuum membrane distillation applied to remediation of dyeing wastewater. *J. Environ. Sci. Heal. Part A* **2019**, *54*, 1337–1347. [CrossRef] [PubMed]

32. Schnittger, J.; McCutcheon, J.; Hoyer, T.; Weyd, M.; Fischer, G.; Puhlfürß, P.; Halisch, M.; Voigt, I.; Lerch, A. Hydrophobic ceramic membranes in MD processes—Impact of material selection and layer characteristics. *J. Membr. Sci.* **2021**, *618*, 118678. [CrossRef]

33. Sparenberg, M.-C.; Hanot, B.; Molina-Fernández, C.; Luis, P. Experimental mass transfer comparison between vacuum and direct contact membrane distillation for the concentration of carbonate solutions. *Sep. Pur. Technol.* **2021**, *275*, 119193. [CrossRef]

34. Guan, G.; Yang, X.; Wang, R.; Field, R.; Fane, A.G. Evaluation of hollow fiber-based direct contact and vacuum membrane distillation systems using aspen process simulation. *J. Membr. Sci.* **2014**, *464*, 127–139. [CrossRef]

35. Mericq, J.-P.; Laborie, S.; Cabassud, C. Evaluation of systems coupling vacuum membrane distillation and solar energy for seawater desalination. *Chem. Eng. J.* **2011**, *166*, 596–606. [CrossRef]

36. Stankiewicz, A.; Moulijn, J.A. Process intensification: Transforming chemical engineering. *Chem. Eng. Prog.* **2000**, *96*, 22–23.

37. Criscuoli, A.; Drioli, E. New metrics for evaluating the performance of membrane operations in the logic of process intensification. *Ind. Eng. Chem. Res.* **2007**, *46*, 2268–2271. [CrossRef]

38. Mengual, J.I.; Khayet, M.; Godino, M.P. Heat and mass transfer in vacuum membrane distillation. *Int. J. Heat Mass Transf.* **2004**, *47*, 865–875. [CrossRef]

Article

Wet Flue Gas Desulphurization (FGD) Wastewater Treatment Using Membrane Distillation

Noah Yakah [1,*], Imtisal-e- Noor [2], Andrew Martin [2,*], Anthony Simons [1] and Mahrokh Samavati [2]

[1]   Department of Mechanical Engineering, University of Mines and Technology, Tarkwa P.O. Box 237, Ghana
[2]   Department of Energy Technology, KTH Royal Institute of Technology, 10044 Stockholm, Sweden
*   Correspondence: nyakah@umat.edu.gh (N.Y.); andrew.martin@energy.kth.se (A.M.)

Abstract: The use of waste incineration with energy recovery is a matured waste-to-energy (WtE) technology. Waste incineration can reduce the volume and mass of municipal solid waste significantly. However, the generation of high volumes of polluting flue gases is one of the major drawbacks of this technology. Acidic gases are constituents in the flue gas stream which are deemed detrimental to the environment. The wet flue gas desulphurization (FGD) method is widely employed to clean acidic gases from flue gas streams, due to its high efficiency. A major setback of the wet FGD technology is the production of wastewater, which must be treated before reuse or release into the environment. Treating the wastewater from the wet FGD presents challenges owing to the high level of contamination of heavy metals and other constituents. Membrane distillation (MD) offers several advantages in this regard, owing to the capture of low-grade heat to drive the process. In this study the wet FGD method is adopted for use in a proposed waste incineration plant located in Ghana. Through a mass and energy flow analysis it was found that MD was well matched to treat the 20 m^3/h of wastewater generated during operation. Thermal performance of the MD system was assessed together with two parametric studies. The thermal efficiency, gained output ratio, and specific energy consumption for the optimized MD system simulated was found to be 64.9%, 2.34 and 966 kWh/m^3, respectively, with a total thermal energy demand of 978.6 kW.

Keywords: waste-to-energy; municipal solid waste; flue gas desulphurization; membrane distillation; thermal performance; thermal efficiency; gained output ratio; specific energy consumption

Citation: Yakah, N.; Noor, I.-e.; Martin, A.; Simons, A.; Samavati, M. Wet Flue Gas Desulphurization (FGD) Wastewater Treatment Using Membrane Distillation. *Energies* **2022**, *15*, 9439. https://doi.org/10.3390/en15249439

Academic Editor: Alessandra Criscuoli

Received: 13 October 2022
Accepted: 9 December 2022
Published: 13 December 2022

Publisher's Note: MDPI stays neutral with regard to jurisdictional claims in published maps and institutional affiliations.

1. Introduction

Waste-to-energy (WtE) technology has been established to be an appropriate method of dealing with municipal solid waste (MSW) worldwide [1]. Most developed countries have embraced the use of it as an attractive means of treating non-recyclable and non-reusable waste because not only it minimizes the risks and environmental concerns associated with disposing copious quantities of MSW into landfill sites, but also it allows production/recovery of useful energy (e.g., electricity). The use of MSW as fuel to generate energy can reduce the over-dependence on fossil fuels as sources of energy.

Waste incineration is the most matured and widely used WtE technology [1–3]. Waste incineration can reduce the volume of MSW by 80 to 95% and the mass by 70 to 75% [4]. The major setback with the use of this technology is the significant amount of pollutants that are produced. There are some constituents (e.g., hydrofluoric (HF), hydrochloric (HCl), and sulphur dioxide (SO_2)) in the flue gases emanating from waste incinerators which are proven to have a detrimental impact on the environment [5]. Therefore, strict limits on the amount of such constituents are set. Different technologies have been considered for removal of such emissions (e.g., wet, semi-dry, and dry scrubbing).

Absorption and adsorption are two distinct processes by which acidic gases are cleaned from flue gas streams emanating from waste incinerators. These processes are classified either as non-regenerative or regenerative. The non-regenerative technologies are further divided into three; dry, semi-dry and wet methods [6,7]. Figure 1 depicts the various methods in non-regenerative flue gas desulphurization (FGD) technologies that are used in acid gas cleaning. Wet scrubbing is a FGD method that is widely employed in cleaning acidic gases in flue gas streams, with Japan, USA and Germany enjoying the most patronage [8]. Advantages of this technology includes higher rates of desulphurization, relative ease of operation, and smaller equipment. The limestone wet method of FGD is reported to be the most widely used technology in the cleaning of acidic gases in flue gas streams from waste incineration plants [9]. A study by Lecomte et al. [10] indicates that removal efficiencies of up to 99% can be achieved using this technology. The major disadvantage with the use of this method is the production of wastewater, which must be treated prior to reuse or release into the environment. There are several wastewater treatment methods that are employed in the treatment of various types of wastewater. Membrane separation methods employed in the treatment of wastewater include microfiltration, ultrafiltration, reverse osmosis (RO), etc. RO has enjoyed the most patronage among the various types of membrane separation techniques. However, drawbacks include higher electricity demand in providing high pressures at the intake, consequently affecting the membrane's long-term performance [11]. There is, therefore, the need to explore other membrane separation processes.

Membrane distillation (MD) is a promising novel technology used in separation processes where only water molecules can pass through a porous hydrophobic membrane material. The application of MD technology is widespread in the desalination of seawater and brackish water [12–15], and the treatment of wastewater that is polluted with radioactive substances [16,17]. MD is also reported to have potential in the treatment of oily wastewater from industries [18–20]. However, relatively few works have been carried out on the application of MD technology in the treatment of flue gas condensate. Chuanfeng et al. [21] investigated the use of MD in the framework of combined heat and power (CHP) plants in Sweden. Subsequently, a pilot unit was installed at the Vattenfall Idb ä cken CHP plant (a biofuel-fired plant) during 2006 and 2007 [22]. Later, a follow-up investigation considered water recovery from flue gas condensate in MSW-fired cogeneration plants using MD [11], and employed both laboratory and pilot-scale air gap MD modules in combination with techno-economic analyses. The aforementioned studies demonstrated that MD offers equal or superior separation efficiency as compared to RO with higher specific costs. The availability of low-grade heat placed an upper limit on the capacity of the MD system to about 100 m^3/h. However, their work was carried out on cogeneration plants, which supply both heat and electricity, and not for WtE facilities operating in condensing mode, where electricity is the only energy service provided. Therefore, there is the need for further investigation if the MD technology is to be integrated in a WtE plant in a tropical country such as Ghana, where the heat demand is low and thus cogeneration is unprofitable when selected. Moreover, to meet the MD energy demand, the possibility of using the recovered waste heat during the cooling of the flue gas stream prior to the particulate matter (PM) separation process is considered.

This study, therefore, mainly focuses on different models of a waste incineration plant simulated using Aspen Plus$^{®}$ software (version 11). In other words, three of the developed models used for the study are presented, with emphasis on the simulation of acid gas cleaning using wet FGD and the treatment of the produced wastewater via MD technology. The current study is a subsequent part of a research investigation with the broader aim of proposing optimal integration of WtE in Ghana.

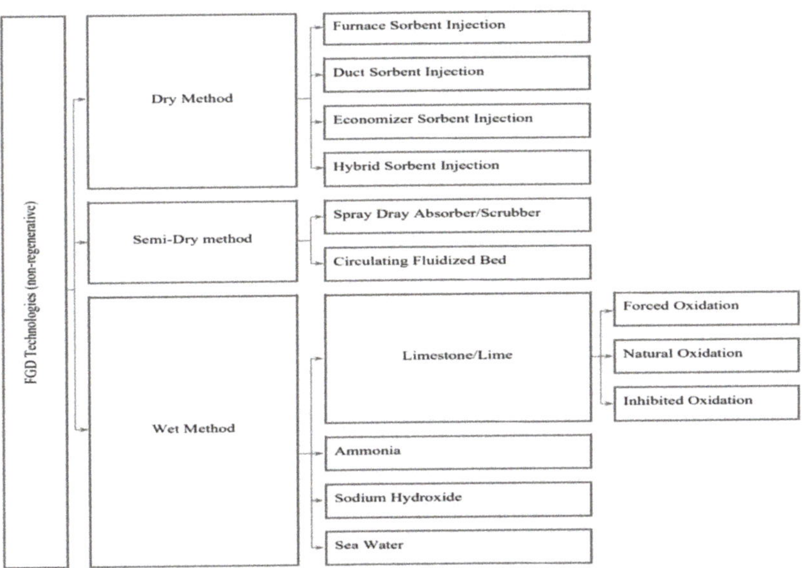

Figure 1. Non-regenerative flue gas desulphurization technologies [23].

2. System Description

This section describes the overall flow chart used in the study, and the system design of a wet scrubbing system and how it is integrated to clean acidic gases in the flue gas stream.

2.1. Block Flowchart of the Incineration Plant Used in the Study

The integrated system block flow chart for the waste incineration plant used in this research is depicted in Figure 2. MSW is first fed into the incinerator/boiler, where sufficient air is added to aid in the complete oxidation of the MSW. After combustion of the MSW in the incinerator/boiler, the flue gas (which carries with it a high energy and particulate matter) and ash are produced. While ash is collected at the bottom, the flue gas stream exits into a heat recovery steam generator (HRSG). After heat exchange with high pressure water to produce high pressure steam, the flue gas is first cooled down further before entering a particulate matter (PM) separation device. At this stage, PM is separated from the flue gas stream using either a cyclone, filter bag, electrostatic precipitator PM separation device, or a combination of these. Then the flue gas stream goes into the wet FGD where water and aqueous calcium carbonate ($CaCO_3$) are added to clean out acidic gases. The flue gas stream is further cleaned before its release to the environment through a stack. The produced wastewater after acid gas cleaning, on the other hand, is sent to the MD system for treatment prior to reuse or disposal into the environment. The wastewater that goes into the MD system is treated and produces a cleaned water (permeate), and the captured solids and the remains in the concentrate (retentate) can be returned to the MD system for further cleaning or disposed.

2.2. Design of a Wet Scrubbing System

A typical design of a limestone wet method of FGD technology has two basic stages. In the first stage, only water (acidic condition) is added. This stage is only effective in the removal of HF, HCl and sulphur trioxide. The removal of SO_2 in the first stage is low due to the presence of HCl, which affects its absorption. Therefore, a second stage is incorporated where a liquid with higher pH (neutral or alkaline condition) is added to remove the SO_2. Studies [6,7,24–26] indicate that the second stage is capable of removing other acidic compounds that are present in the flue gas stream.

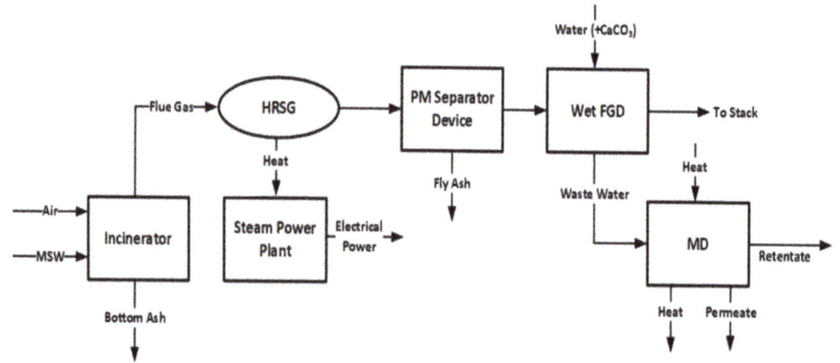

Figure 2. Block flowchart of incineration plant used for the study.

3. Methodology

The integrated system is divided into three subsystems that are simulated using Aspen Plus®. This section describes these three subsystem models simulated.

The first is the waste incineration plant, which is to determine the volume of emissions, including the parameters of the various constituents, that were generated after the combustion of the MSW. The second model is the wet scrubbing process, employed to determine the volume of water needed to clean acid gases (limited to only HCl and SO_2 in this study) from the flue gas stream and subsequently the volume of flue gas condensate (wastewater) generated in the process. The third model is the MD technology which is used to treat the flue gas condensate produced during the acid gas cleaning.

The relations used in performing the thermal analysis of the MD system are also presented in this section.

3.1. Models Used in the Study

3.1.1. Model of the Waste Incineration Plant

The waste incineration plant Aspen Plus® flowsheet [27] is shown in Figure 3. The simulated plant has a nominal incineration capacity of 240 metric tons of MSW per day, generating 30 MW of electrical power with an efficiency of approximately 31%. The waste incineration plant model has four stages: (1) drying of the MSW, (2) combustion of the MSW, (3) steam generation for only electricity generation, i.e., condensing mode and (4) PM separation.

In the waste incineration plant, the wet MSW (WET-MSW) is sent into a vessel (DRY-REAC), where hot air (HOTAIR) is mixed with WET-MSW. A calculator block is defined in Aspen Plus to control the drying process in another vessel (DRY-FLSH) and the by-products from this vessel are a dry MSW (DRY-MSW) and an exhaust vapour (EXHAUST), which is discharged into the atmosphere.

DRY-MSW is now ready to be combusted. As its composition can vary based on the source and regional factors (e.g., topography, seasons, food habits etc.), it has been defined as non-conventional in the model. Consequently, for successful simulation of combustion process, DRY-MSW first needs to be defined based on its content. Therefore, an extra vessel (DECOMP) is included in the flowsheet where DRY-MSW is broken down into its various elemental constituents (Q-DECOMPOST). Q-DECOMPOST is then sent into the combustion chamber (BURN), where sufficient air (ATM-AIR) is added to achieve a complete oxidation of the MSW. Energy is recovered from the flue gases (CPROD-H) from the combustion process in the heat exchanger (HRSG) for the generation of superheated steam (HPSTEAM), which turns a steam turbine (ST-TURB) for the generation of electrical power (WT-TURB).

After the recovery of heat energy from CPROD-H, there is a drop in temperature in the flue gas (CPROD-C) before entering particulate matter (PM) separation devices. In the model depicted in Figure 3, all three types (cyclone, bag filter and the electrostatic precipitator (ESP)) of PM removal devices are incorporated. The flue gas stream after the PM separation (ESP-GAS) then goes into the wet scrubbing system for cleaning of acidic gases.

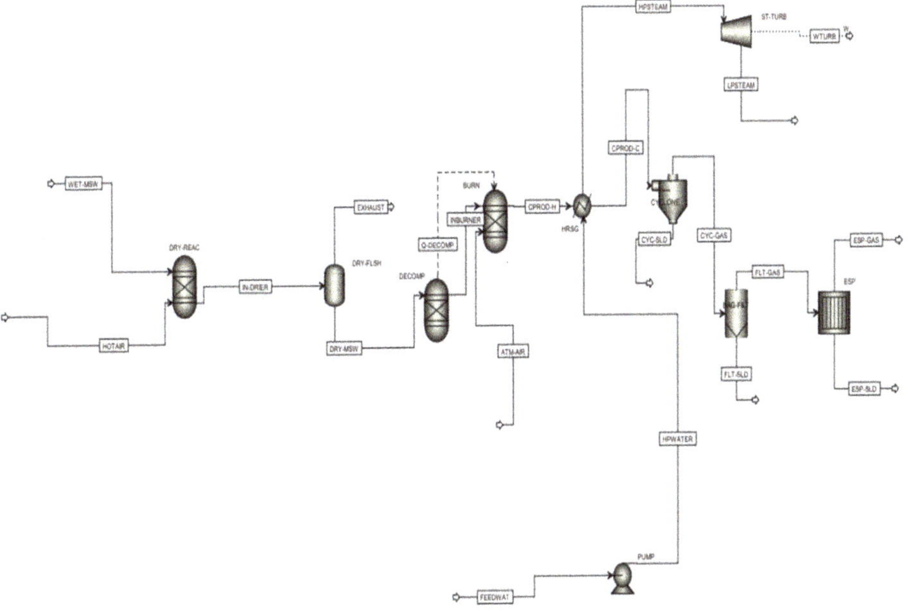

Figure 3. Waste Incineration Aspen Plus® model flowsheet [27].

3.1.2. Model of the Wet Scrubbing Process

An Aspen Plus® model (Figure 4) of a wet scrubbing process was developed in order to simulate the cleaning process of acid gases from produced flue gas in the previous model. The main output of this model is to determine the amount of flue gas condensate (wastewater) that would be generated. The base method used in Aspen Plus® is ELECNRTL.

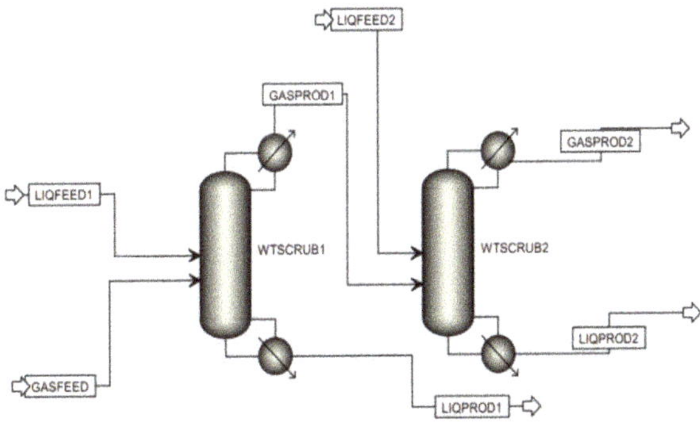

Figure 4. Wet scrubbing process Aspen Plus® flowsheet [27].

As can be seen in Figure 4, the developed model has two stages. RadFrac (WTSCRUB1) from the Aspen block built-in library, which is the acidic scrubber, is selected where the flue gas stream from the waste incinerator (GASFEED) and water (LIQFEED1) are the feed streams. The products then are wastewater (LIQPROD1), and the flue gas stream (GASPROD1). The second wet scrubber (WTSCRUB2), selected as the same as the first scrubber in Aspen Plus, is the alkaline (or neutral) scrubber with two feeds: the partially cleaned flue gas (GASPROD1) from WTSCRUB1 and a liquid feed (LIQFEED2). The latter is an alkaline solution, which in this study is considered to be calcium carbonate ($CaCO_3$).

Table 1 lists the operating condition used in the modelling of the wet scrubbing process. It is assumed that there is no temperature and pressure drop in GASPROD1 from WTSCRUB1 to WTSCRUB2.

Table 1. Operating conditions in the wet scrubbing process.

Parameter	Value
Flue gas feed temperature to WTSCRUB1	160 °C
Flue gas feed pressure to WTSCRUB1	1.01 bar
Liquid feed temperature to WTSCRUB1	30 °C
Liquid feed pressure to WTCSCRUB1	1.5 bar
Operating Pressure in WTSCRUB1	1.01 bar
Number of stages in WTSCRUB1	10
Liquid feed temperature to WTSCRUB2	35 °C
Liquid feed pressure to WTSCRUB2	1.5 bar
Number of stages in WTSCRUB2	10

3.1.3. Model of the MD System

There are no available blocks in the Aspen Plus built-in library that can readily be used in the simulation of an MD unit. Hence, it was modelled using a customised USER Model2 in Aspen Plus. An excel file sheet built into Aspen Plus was modified with data obtained from the simulation of the wet scrubbing model. Table 2 is a list of the operating parameters used in the simulation of the MD system. This simulation work is based on the MD system presented by Imtisal-e-Noor et al. [28]. However, there are a few differences between that simulation work and this current work. The current model has a single air gap MD module relative to the dual-cascaded MD modules used in that work. There are also differences in parameters, such as the feed inlet temperature, coolant inlet temperature, density of the feed, and the composition of the flue gas condensate. The feed and coolant inlet pressures, however, remain the same.

Table 2. Operational parameters of MD system.

Operating Parameter	Value
Feed flowrate	1500 L/h
Feed inlet temperature	85 °C
Feed inlet pressure	1.0 bar
Coolant flowrate	1500 L/h
Coolant inlet temperature	26 °C
Coolant inlet pressure	1.0 bar

The Aspen Plus® MD model flowsheet is shown in Figure 5, and the base method used for this model is IDEAL. The flue gas condensate or wastewater (WWSCRUB) generated from the wet scrubbing process is collected into a tank (TNK) at a temperature of 56.7 °C. The wastewater stored in the tank (FD1) is then passed through a heat exchanger (HX) and heated up to a temperature of 85 °C using heat from the cooling of the flue gas stream (from 440 °C to 160 °C) before particulate matter separation. The heated flue gas condensate (FD2) then goes into the MD module (MD). The temperature of the flue gas stream drops because of the latent heat of vaporization which corresponds to the permeate flux passing through the membrane. The concentrate and permeate streams from the MD module are referred to as RET and PERM, respectively. The treated water (PERM) is then collected into another tank for reuse.

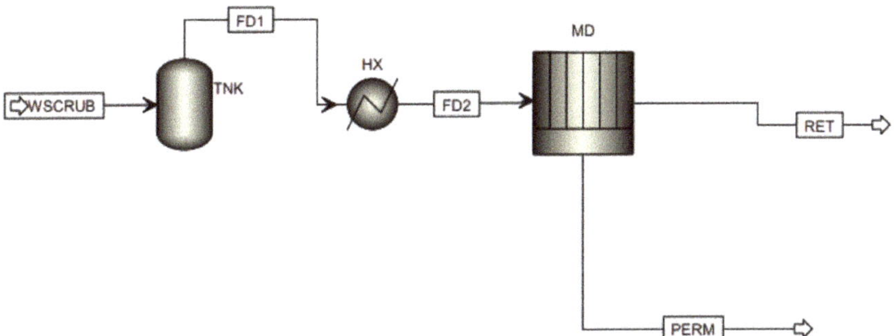

Figure 5. MD system Aspen Plus® flowsheet [27].

During the simulation of the MD system, the coolant stream is considered to be water to be pumped from the surroundings (e.g., from a river) with an ambient temperature of between 25 to 27 °C (the ambient temperature of water in Ghana).

3.2. Thermal Energy Analysis of MD Systems

The permeate flux, or simply flux, of an MD (J_p) is considered the most relevant metric used in the assessment of any membrane technology. It is defined as the flow rate of permeate flowing through the membrane measured in kg/m^2 s and can be expressed mathematically as [29]

$$J_p = \frac{\dot{m}_p}{A} \tag{1}$$

where $\dot{m}_p$ is the mass flow rate of the permeate measured in kg/s and A is the effective membrane area measured in m^2.

The thermal efficiency (TE) or evaporative thermal efficiency of MD systems is defined as the ratio of the latent heat of vaporization to the total (latent and conduction) heat. The TE of MD systems is considered an effective tool in the measurement of desired thermal transport. It can be expressed mathematically as [29]

$$TE\ (\%) = \frac{\dot{m}_p\ \Delta H_{v,w}}{Q_m} \times 100 \tag{2}$$

where $\Delta H_{v,w}$ refers to the enthalpy of vaporization of the water in kJ/kg, and Q_m is the total heat flux through the membrane in kW, which can be determined by using the relation in Equation (3).

$$Q_m = \dot{m}_f C_p \left(T_{f,in} - T_{f,out} \right) \tag{3}$$

In this expression, $\dot{m}_f$ refers to the feed mass flow rate measured in kg/s, C_p refers to the feed water specific heat measured in kJ/kg °C, while $T_{f,in}$ and $T_{f,out}$ refer to the inlet and outlet feed water temperatures, respectively, measured in °C.

Another important parameter for evaluation of the thermal performance of an MD system is specific energy consumption (SEC). It is defined as the energy required to produce 1 m^3 of distillate water in MD systems and can be determined using the following relation [30]

$$SEC \left(\frac{\text{kWh}}{\text{m}^3} \right) = \left[\frac{Q_m \rho}{J_p\ A} \right] / 3600 \tag{4}$$

where ρ refers to the density of water measured in kg/m^3.

Gained output ratio (GOR) is defined as the ratio of thermal energy that is required to produce distillate water in an MD system to the energy input to the system. GOR, a dimensionless parameter, can be expressed mathematically using the following equation [31]

$$GOR = \frac{J_p A \triangle H_{v,w}}{E_{in}} \tag{5}$$

where E_{in} refers to the total power input to the system measured in kW.

4. Results and Discussion

In this section, results for both the wet scrubbing and MD processes (including a parametric analysis on the MD system) are presented and discussed.

4.1. Results from the Wet Scrubbing Process

Table 3 presents some key simulation results of the wet scrubbing process. The temperature of the flue gas condensate was 56.7 °C, at a pressure 1.01 bar. The temperature of the flue gas condensate was above its dew temperature. The temperature of the cleaned flue gas (after acid gas cleaning) which continues into the stack to be emitted into the atmosphere is 52.2 °C at a pressure of 1.01325 bar and this is also above its dew point. In all the process can achieve an overall cleaning efficiency of over 99% for SO_2 and over 95% for HCl. There was also a little reduction in the other constituents of the flue gases exiting the wet scrubber at the end of the cleaning process. The total flow rate of liquid feed used in the wet scrubbing was 70,000 kg/h of water, 35,000 kg/h in the first scrubber and 35,000 kg/h in the second scrubber. It must be noted that in the second scrubber the liquid feed is a mixture of water and $CaCO_3$. (The mole fraction ratio of water to $CaCO_3$ is 0.9:0.1).

Table 3. Simulation results from the wet scrubbing process.

Parameter	Value
Flue gas condensate (wastewater) temp	56.7 °C
Flue gas condensate (wastewater) pressure	1.01 bar
Cleaned gas temperature	52.2 °C
Cleaned gas pressure	1.01 bar
SO_2 cleaning efficiency	Over 99%
HCl cleaning efficiency	Over 95%
Volumetric flow rate of wastewater	19.4 m^3/h

4.2. MD Model Simulation Results

This section presents the results obtained from the simulation of the optimized MD model used in the study. The thermal performance of the MD system is also presented and discussed. Table 4 list the results obtained after simulation of the MD model.

Table 4. Simulation results from the MD system.

Parameter	Value
Feed/concentrate outlet temperature	77.1 °C
Coolant stream outlet temperature	34.2 °C
Membrane flux	6.22 L/m^2/h
Total thermal energy demand	978.6 kW
Membrane area	699 m^2
Number of modules	303
GOR	2.34
SEC	966 kWh/m^3
TE	64.9%

4.2.1. Thermal Efficiency

The thermal efficiency of the MD system was determined using Equation (2) when increasing the feed inlet temperature from 75 °C to 90 °C while maintaining the coolant inlet temperature at 26 °C. Values of the determined thermal efficiency against the corresponding feed inlet temperature are reported in Figure 6a. It can be observed in the figure that increasing the feed inlet temperature from 75 °C to 90 °C resulted in an increment of the TE from 50.7% to approximately 73%. This increment in TE can be attributed to the increase in permeation when the feed/concentrate inlet temperature increases. The thermal efficiency of the MD system was determined again using Equation (2), but this time increasing the coolant inlet temperature from 15 °C to 32 °C while maintaining the feed inlet temperature at 85 °C. Values of the determined thermal efficiency against the corresponding coolant inlet temperature are reported in Figure 6b. It can be observed that increasing the coolant inlet temperature from 15 °C to 32 °C decreases the TE from 75.4% to approximately 57%. This decrease in TE can be attributed to the low level of permeation due to the increase in the coolant inlet temperature. These results conform to the results obtained in research by Shahu et al. and Elmarghany et al. [32,33].

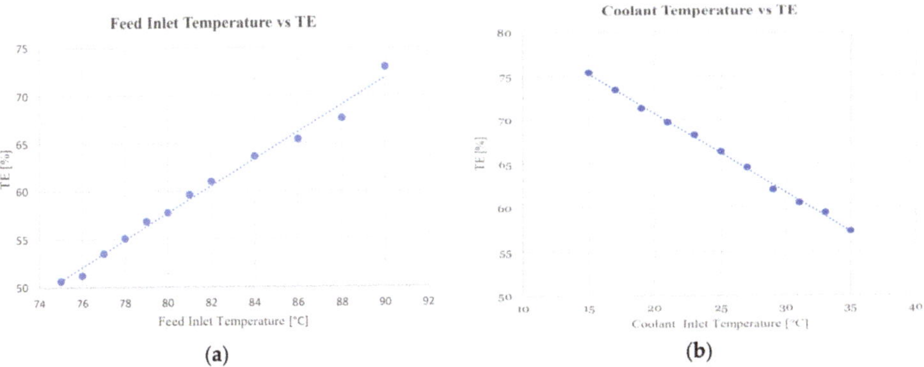

Figure 6. Effects on increasing feed/coolant inlet temperature versus TE: (**a**) Effects on increasing feed inlet temperature vs. resulting SEC. (**b**) Effects on increasing coolant inlet temperature vs. resulting TE.

4.2.2. GOR

The GOR of the MD system was determined using Equation (3) by increasing the feed inlet temperature from 75 °C to 90 °C while maintaining the coolant temperature inlet temperature at 26 °C. The values of the determined GOR against the feed inlet temperature are reported in Figure 7a. It can be observed from that increasing the feed inlet temperature from 75 °C to 90 °C increases the GOR from 1.82 to 2.53. Increasing the feed inlet temperature increases permeation, which in turn increases the driving force of the permeate, as much less thermal energy is required to produce distillate water in the MD system. The GOR of the MD system was determined again using Equation (5), but this time increasing the coolant inlet temperature from 15 °C to 32 °C while maintaining the feed inlet temperature at 85 °C. Values of the GOR against the corresponding inlet temperature are reported in Figure 7b. It can be observed from the figure that increasing the coolant inlet temperature from 15 °C to 32 °C decreases the GOR from 2.72 to 2.06. This decrease can be attributed to the fact that increasing the coolant inlet temperature decreases the level of permeation and the driving force; therefore, higher amount of thermal energy is required to produce distillate water in the MD system. These results conform to the results obtained in research works carried out by Shahu et al. and Elmarghany et al. [32,33].

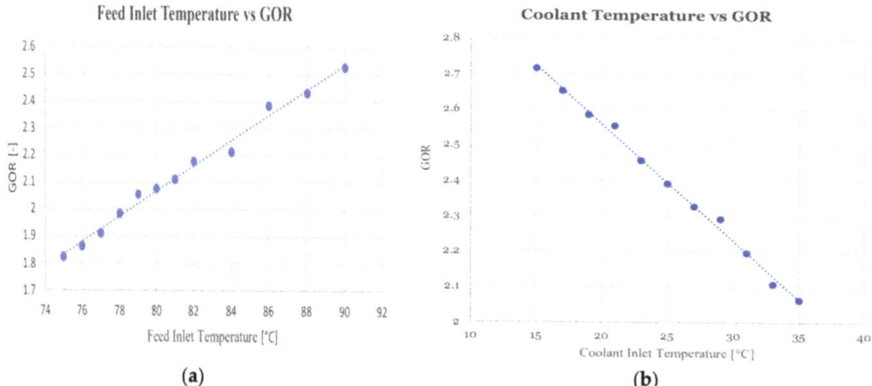

Figure 7. Effects on increasing feed/coolant inlet temperature versus GOR: (**a**) Effects on increasing feed inlet temperature vs. resulting GOR. (**b**) Effects on increasing coolant inlet temperature vs. resulting GOR.

4.2.3. SEC

The SEC of the MD system was determined using Equation (3) by increasing the feed inlet temperature from 75 °C to 90 °C while maintaining the coolant temperature inlet temperature at 26 °C. The values of the determined SEC against the corresponding feed inlet temperature are reported in Figure 8a. It can be observed that increasing the feed inlet temperature from 75 °C to 90 °C decreases the SEC from 1237.4 kWh/m^3 to approximately 860 kWh/m^3. As mentioned earlier, increasing the feed inlet temperature increases permeation and thus increases the driving force of the permeate which in turn decreases the energy required to produce the distillate water, therefore decreasing the SEC. The SEC of the MD system was determined again using Equation (4), but this time increasing the coolant inlet temperature from 15 °C to 32 °C while maintaining the feed inlet temperature at 85 °C. The values of the determined SEC against the corresponding inlet temperature are reported in Figure 8b. It can be observed that increasing the coolant from 15 °C to 32 °C increases the SEC from 831.2 kWh/m^3 to approximately 1074 kWh/m^3. Increasing the coolant inlet temperature decreases the level of permeation which in turn decreases the driving force of the permeate, therefore more energy is needed to produce the distillate water. These results conform to the results obtained in research works carried out by Shahu et al. and Elmarghany et al. [32,33].

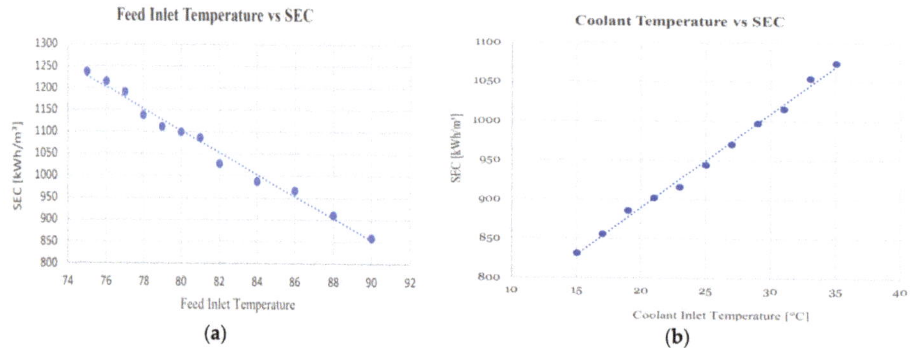

Figure 8. Effects on increasing Feed/Coolant Inlet temperature versus SEC: (**a**) Effects on increasing feed inlet temperature vs. resulting SEC. (**b**) Effects on increasing coolant inlet temperature vs. resulting SEC.

5. Conclusions

The results obtained from the simulation of the MD process indicates it is possible to achieve almost 100% separation of the various constituents of the flue gas condensate. Additionally, a thermal performance assessment of the MD process indicates that increasing the feed/concentrate inlet while minimizing that of the coolant inlet temperature is the optimum means of operating the MD. It is therefore imperative to state that the coolant inlet temperature of 25 °C is reasonable and the feed inlet temperature should be at 90 °C.

Results obtained earlier from the simulation of the wet FGD model indicate almost 100% cleaning of acid gases (HCl and SO_2) from the flue gas stream, where a total volumetric flow rate of 20 m^3/h was used in the process. The total flue gas condensate produced at the end of the process was approximately 19.44 m^3/h.

In conclusion, it is worth noting that the energy recovered during the cooling of the flue gas stream prior to PM separation is adequate for the operation of the integrated MD system, and the wet FGD technology is an effective method of cleaning acid gases from flue gas streams (achieving separation efficiencies of over 95% for the HCl and over 99% for the SO_2). In addition, the MD technology is an effective method that can be used in the separation of these acid gases in the flue gas condensate produced in the wet FGD technology.

It is recommended that a techno-economics analysis on both the wet FGD and MD technologies are performed on a typical waste incineration plant with energy recovery, with Ghana as the location for the operation of the plant.

Author Contributions: Conceptualization, N.Y. and A.M.; methodology, N.Y. and A.M.; software, N.Y., M.S, and I.-e.N.; validation, N.Y.; formal analysis, N.Y.; writing—original draft preparation, N.Y.; writing—review and editing, N.Y., M.S., I.-e.N., A.M., and A.S.; supervision, A.M., and A.S.; All authors have read and agreed to the published version of the manuscript.

Funding: This research received no external funding. The APC was funded by a discount voucher received by A.M.

Data Availability Statement: Not applicable.

Acknowledgments: Authors wish to thank University of Mines and Technology, (UMaT), Tarkwa, Ghana and Royal Institute of Technology (KTH), Stockholm, Sweden for their support.

Conflicts of Interest: The authors declare no conflict of interest.

References

1. Dong, J.; Tang, Y.; Nzihou, A.; Chi, Y.; Weiss-Hortala, E.; Ni, M.; Zhou, Z. Comparison of waste-to-energy technologies of gasification and incineration using life cycle assessment: Case studies in Finland, France and China. *J. Clean. Prod.* **2018**, *203*, 287–300, ISSN 0959-6526. [CrossRef]
2. Demirbas, A. Combustion Systems for Biomass Fuel. *Energy Sources Part A Recovery Util. Environ. Eff.* **2007**, *29*, 303–312. [CrossRef]
3. Lombardi, L.; Carnevale, E.; Corti, A. A review of technologies and performances of thermal treatment systems for energy recovery from waste. *Waste Manag.* **2015**, *37*, 26–44. [CrossRef]
4. The World Bank. *Municipal Solid Waste Incineration: Technical Guidance Report*; *Technical Report*; The International Bank for Reconstruction and Development: Washington, DC, USA, 1999. Available online: https://goo.gl/kABenL (accessed on 20 October 2020).
5. Mohajan, H. Acid Rain is a Local Environment Pollution but Global Concern. *Open Sci. J. Anal. Chem.* **2019**, *3*, 47–55.
6. Neuwahl, F.; Cusano, G.; Benavides, J.G.; Holbrook, S.; Roudier, S. *Best Available Techniques (BAT) Reference Document for Waste Incineration*; *Report*; Publication Office of the European Union: Luxembourg, 2019. [CrossRef]
7. Johnke, G. Draft of a German Report with Basic Informations for a BREF-Document "Waste Incineration"; Report. 2001. Available online: http://files.gamta.lt/aaa/Tipk/tipk/4_kiti%20GPGB/63.pdf (accessed on 13 December 2021).
8. Zhao, G. Theoretical Study for Sulfur Dioxide Absorption on Limestone Wet Flue Gas Desulphurization. Master's Thesis, Northeastern University, Shenyang, China, 2008.
9. Carletti, C.; Bjondahl, F.; De Blasio, C.; Ahlbeck, J.; Järvinen, L.; Westerlund, T. Modeling limestone reactivity and sizing the dissolution tank in wet flue gas desulfurization scrubbers. *Environ. Prog. Sustain. Energy* **2013**, *32*, 663–672. [CrossRef]

10. Lecomte, T.; Ferreria De La Fuente, J.; Neuwahl, F.; Canova, M.; Pinasseau, A.; Jankov, I.; Brinkmann, T.; Roudier, S.; Delgado Sancho, L. *Best Available Techniques (BAT) Reference Document for Large Combustion Plants. Industrial Emissions Directive 2010/75/EU (Integrated Pollution Prevention and Control); EUR 28836 EN*; Publications Office of the European Union: Luxembourg, 2017; ISBN 978-92-79-74303-0. [CrossRef]

11. Noor, I.-e.; Martin, A.; Dahl, O. Water recovery from flue gas condensate in municipal solid waste fired cogeneration plants using membrane distillation. *Chem. Eng. J.* **2020**, *399*, 125707. [CrossRef]

12. Alkhudhiri, A.; Darwish, N.; Hilal, N. Membrane distillation: A comprehensive review. *Desalination* **2012**, *287*, 2–18. [CrossRef]

13. Camacho, L.M.; Dumée, L.; Zhang, J.; Li, J.-d.; Duke, M.; Gomez, J.; Gray, S. Advances in Membrane Distillation for Water Desalination and Purification Applications. *Water* **2013**, *5*, 94–196. [CrossRef]

14. Qtaishat, M.R.; Banat, F. Desalination by solar powered membrane distillation systems. *Desalination* **2013**, *308*, 186–197. [CrossRef]

15. Rahaoui, K.; Ding, L.C.; Tan, L.P.; Mediouri, W.; Mahmoudia, F.; Nakoa, K.; Akbarzadeh, A. Sustainable Membrane Distillation Coupled with Solar Pond. *Energy Procedia* **2017**, *110*, 414–419. [CrossRef]

16. Liu, H.; Wang, J. Treatment of radioactive wastewater using direct contact membrane distillation. *J. Hazard. Mater.* **2013**, *261*, 307–315. [CrossRef]

17. Korolkov, I.V.; Yeszhanov, A.B.; Zdorovets, M.V.; Gorin, Y.G.; Güven, O.; Dosnagambetova, S.S.; Khlebnikov, N.A.; Serkv, K.V.; Krasnopyorova, M.V.; Milts, O.S.; et al. Modification of PET ion track membranes for membrane distillation of low-level liquid radioactive wastes and salt solutions. *Sep. Purif. Technol.* **2019**, *227*, 115694. [CrossRef]

18. Ricceri, F.; Giagnorio, M.; Farinelli, G.; Blandini, G.; Minella, M.; Vione, D.; Tiraferri, A. Desalination of Produced Water by Membrane Distillation: Effect of the Feed Components and of a Pre-treatment by Fenton Oxidation. *Sci. Rep.* **2019**, *9*, 14964. [CrossRef]

19. Tavakkoli, S.; Lokare, O.; Vidic, R.; Khanna, V. Shale gas produced water management using membrane distillation: An optimization-based approach. *Resour. Conserv. Recycl.* **2020**, *158*, 104803. [CrossRef]

20. Said, I.A.; Chomiak, T.R.; He, Z.; Li, Q. Low-cost high-efficiency solar membrane distillation for treatment of oil produced waters. *Sep. Purif. Technol.* **2020**, *250*, 117170. [CrossRef]

21. Chuanfeng, L.; Martin, A. *Membrane Distillation and Applications for Water Purification in Thermal Cogeneration—A Pre-study*; Varmeforsk Service AB: Stockholm, Sweden, 2005; ISSN 0282-3772.

22. Kullab, A. Desalination using Membrane Distillation: Experimental and Numerical Study. Ph.D. Thesis, KTH Royal Institute of Technology, Stockholm, Sweden, 2011.

23. Jurczyk, M.; Mikus, M.; Dziedzic, K. *Flue Gas Cleaning in Municipal Waste-to-Energy Plants—Part II; Nr. IV/2/2016*; Polska Akademia Nauk Oddział w Krakowie, Komisja Technicznej Infrastruktury Wsi: Krakow, Poland, 2016; pp. 1309–1321. Available online: http://dx.medra.org/10.14597/infraeco.2016.4.2.096 (accessed on 20 November 2021).

24. Hellman, M. Handbok i Vattenkemi för energianläggningar; Energiforsk, Stockholm, Sweden. 2015. Available online: https://energiforsk.se/program/branslebaserad-el-och-varmeproduktion-sebra/rapporter/handbookk-for-vattenkemi-for-energianlaggningar/ (accessed on 23 October 2021).

25. Goldschmidt, H.O.B.; Carlström, H. *Benchmarking of Flue Gas Condensate Cleaning Technologies in Waste-to-Energy Plants*; Energiforsk: Stockholm, Sweden, 2011. Available online: https://energiforskmedia.blob.core.windows.net/media/17905/teknikval-vid-rening-av-roekgaskondensat-i-avfallsfoerbraenningsanlaeggningar-vaermeforskrapport-1184.pdf (accessed on 6 December 2021).

26. Matthews, C. Flue gas desulphurization—Total design. In *Case Studies in Engineering Design*; Matthews, C., Ed.; Butterworth-Heinemann: London, UK, 1998; pp. 217–231. [CrossRef]

27. Solids | Process Modeling | Aspentech. Available online: https://www.aspentech.com/en/products/pages/solids (accessed on 1 November 2021).

28. Noor, I.-e.; Samavati, M.; Martin, A. Evaluation of membrane distillation system for flue gas condensate cleaning. In Proceedings of the ECOS 2020—The 33rd International Conference on Efficiency, Cost, Optimization, Simulation and Environmental Impact of Energy Systems, Osaka, Japan, 29 June–3 July 2020.

29. Swaminathan, J.; Chung, H.W.; Warsinger, D.M.; Lienhard, V.J.H. Energy efficiency of membrane distillation up to high salinity: Evaluating critical system size and optimal membrane thickness. *Appl. Energy* **2018**, *211*, 715–734. [CrossRef]

30. Soomro, M.L.; Kim, W.S. Parabolic-trough plant integrated with direct-contact membrane distillation system: Concept, simulation, performance, and economic evaluation. *Sol. Energy* **2018**, *173*, 348–361. [CrossRef]

31. Khayet, M. Solar desalination by membrane distillation: Dispersion in energy consumption analysis and water production costs (a review). *Desalination* **2013**, *308*, 89–101. [CrossRef]

32. Shahu, V.T.; Thombre, S.B. Theoretical analysis and parametric investigation of an innovative helical air gap membrane desalination system. *Appl. Water Sci.* **2022**, *12*, 18. [CrossRef]

33. Elmarghany, M.R.; El-Shazly, A.H.; Salem, M.S.; Sabry, M.N.; Nady, N. Thermal analysis evaluation of direct contact membrane distillation system. *Case Stud. Therm. Eng.* **2019**, *13*, 100377. [CrossRef]

Article

Osmotic Membrane Distillation Crystallization of NaHCO$_3$

Mar Garcia Alvarez [1,2,*], Vida Sang Sefidi [1,2,*], Marine Beguin [1], Alexandre Collet [1], Raul Bahamonde Soria [1,3] and Patricia Luis [1,2,*]

[1] Materials & Process Engineering (iMMC-IMAP), UCLouvain, Place Sainte Barbe 2, B-1348 Louvain-la-Neuve, Belgium; beguinmarine@hotmail.com (M.B.); collet.alexandre@hotmail.com (A.C.); rabahamonde@uce.edu.ec (R.B.S.)
[2] Research & Innovation Centre for Process Engineering (ReCIPE), Place Sainte Barbe, 2 bte L5.02.02, B-1348 Louvain-la-Neuve, Belgium
[3] Renewable Energy Laboratory, Chemical Sciences Faculty, Universidad Central del Ecuador, Quito 170129, Ecuador
* Correspondence: mar.garcia@uclouvain.be (M.G.A.); vida.sangsefidi@uclouvain.be (V.S.S.); patricia.luis@uclouvain.be (P.L.)

Abstract: A new crystallization process for sodium bicarbonate (NaHCO$_3$) was studied, proposing the use of osmotic membrane distillation crystallization. Crystallization takes place due to the saturation of the feed solution after water evaporation on the feed side, permeating through the membrane pores to the osmotic side. The process operational parameters, i.e., feed and osmotic velocities, feed concentration, and temperature were studied to determine the optimal operating conditions. Regarding the feed and osmotic velocities, values of 0.038 and 0.0101 m/s, respectively, showed the highest transmembrane flux, i.e., 4.4×10^{-8} m^3/m^2·s. Moreover, study of the temperature variation illustrated that higher temperatures have a positive effect on the size and purity of the obtained crystals. The purity of the crystals obtained varied from 96.4 to 100% In addition, the flux changed from 2×10^{-8} to 7×10^{-8} m^3/m^2·s with an increase in temperature from 15 to 40 °C. However, due to heat exchange between the feed and the osmotic solutions, the energy loss in osmotic membrane distillation crystallization is higher at higher temperatures.

Keywords: NaHCO$_3$; osmotic membrane distillation crystallization; membrane contactor

Citation: Garcia Alvarez, M.; Sang Sefidi, V.; Beguin, M.; Collet, A.; Bahamonde Soria, R.; Luis, P. Osmotic Membrane Distillation Crystallization of NaHCO$_3$. *Energies* **2022**, *15*, 2682. https://doi.org/10.3390/en15072682

Academic Editor: Alessandra Criscuoli

Received: 28 February 2022
Accepted: 28 March 2022
Published: 6 April 2022

Publisher's Note: MDPI stays neutral with regard to jurisdictional claims in published maps and institutional affiliations.

1. Introduction

Climate change is redirecting global objectives to regulate greenhouse gas emissions. Industry accounts for 21% of these emissions [1] Thus, in order to minimize waste production and its effect on the environment, the design of more efficient processes is required. Many industrial sectors are already focused on lower energy consumption, such as the pharmaceutical industry, food industry, fine chemicals industry, and construction. However, some of their processes are still far from sustainable [2,3]. That is the case for crystallization, a separation technique for producing or purifying solid products from a supersaturated solution. Crystals have high stability, are easy to store, and have a long life. For these reasons, there is an immense requirement for their production from industry [4].

On a larger scale, several principles are used to form crystals, such as cooling of the feed solution, evaporation of the solvent, and anti-solvent techniques. The conventional equipment for performing crystallization is a batch stirred tank, which has several drawbacks. Firstly, the conventional crystallizer cannot provide crystalline solid products of sufficient morphological quality (size, shape, and crystal size distribution), structure (polymorphism), and purity [5]. Secondly, there are some reproducibility issues such as imperfect mixing, where the solution is not homogeneous, and the supersaturation control is limited. Moreover, the points at which crystallization can be performed vary from one batch to another. Furthermore, a great deal of energy is needed either to heat/cool the solution in a conventional evaporator or to power vacuum systems, which are not efficient [6].

In addition, the stirred tank mostly operates as a batch reactor, meaning that the process is not continuous and has to be stopped to recover the products. It would be more convenient and energetically more efficient to use a continuous process [5,7–9]. As conventional crystallizers have many inconveniences, research has been conducted to find alternatives allowing better control and performance during the crystallization process, and membrane distillation crystallization is one of these alternatives [4].

Osmotic membrane distillation crystallization (OMDC) is an innovative technique in which two liquids are brought into contact through a non-selective hydrophobic microporous membrane [10]. Because the concentration is not the same on both sides, this induces a water activity difference and leads to the evaporation of water from the feed to the osmotic side. Thus, the driving force is the vapor pressure gradient created by the water activity difference between the two sides of the membrane. Figure 1 depicts the mass transfer profile for the OMDC system [7,8].

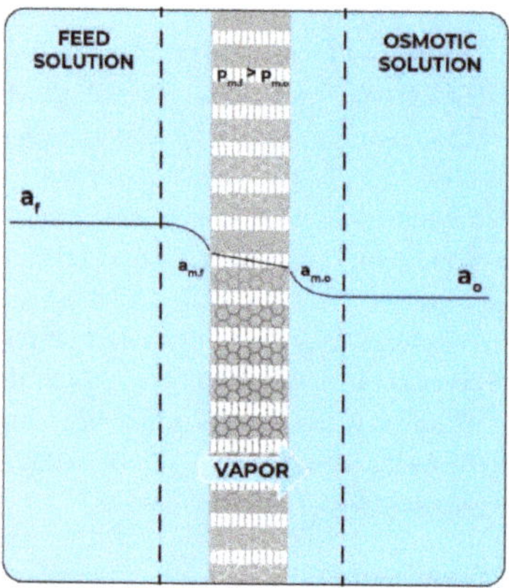

Figure 1. Concentration profiles in osmotic membrane distillation crystallization.

OMDC has advantages over conventional distillation and crystallization processes. This technique has a very high specific contact area, promoting higher mass transfer with more compact equipment than in conventional crystallization or distillation. The main advantage of OMDC is lower energy consumption [11,12]. As the driving force is created through the partial pressure gradient, no additional pressure is required, which allows equipment costs to be reduced and process safety to be increased in comparison with pressure-driven processes. Residual heat or renewable energy can also be used, if available, which could reduce the overall cost and environmental impact [12–14]. Another benefit of OMDC is the use of polymer materials in the equipment, which decreases or even avoids erosion problems [4–7].

OMDC is presented as an alternative option for crystallizing sodium bicarbonate ($NaHCO_3$). $NaHCO_3$ is a salt obtained from a reaction between soda ash and carbon dioxide (CO_2) [15,16]. $NaHCO_3$ is used in various industries such as food, pharmaceuticals, agriculture, etc. However, the purity and the morphology of the obtained crystals play an important role in where the $NaHCO_3$ salts can be used. To crystallize $NaHCO_3$ in a conventional crystallizer, CO_2 must be introduced to the tank atmosphere as $NaHCO_3$, which can easily be converted to CO_2 by heat or stirring [15]. Shifeng Jiang also studied the crystallization of $NaHCO_3$ using a cooling crystallizer to generate more $NaHCO_3$ crystals [17]. However, when using OMDC technology to crystallize $NaHCO_3$, there is no need for the constant addition of CO_2. Moreover, as the solution is not heated, less $NaHCO_3$ is converted to CO_2, which is the main advantage of OMDC for crystallizing $NaHCO_3$.

To the best of our knowledge, no studies have been performed on the crystallization of $NaHCO_3$ using membrane distillation crystallization. However, OMDC has been used for other materials. Israel Ruiz Salmon et al. studied OMDC for the crystallization of sodium carbonate. It was observed that in OMDC, the main resistance was the membrane itself, and the process suffered from concentration polarization and possible wetting [18].

In this study, the main objective was to optimize the OMDC system for the crystallization of $NaHCO_3$. Several operational parameters such as the feed and osmotic velocities, the effect of feed concentration, and the feed temperature were studied. Moreover, the purity, shape, and size of the crystals were analyzed using X-ray diffraction (XRD) and scanning electron microscopy (SEM).

2. Materials and Methods

2.1. Chemicals

The feed solution for each experiment was produced by dissolving $NaHCO_3$ salt (sodium bicarbonate, $\geq$99.7%, AnalaR NORMAPUR, Leuven, Belgium) in ultrapure water, and the osmotic solution was obtained by dissolving sodium chloride (NaCl) (sodium chloride, $\geq$99.9%, AnalaR NORMAPUR, Leuven, Belgium) up to the maximum solubility in ultrapure water.

2.2. Equipment

Figure 2 shows the scheme for the distillation/crystallization setup. The membrane contactor used to carry out experiments was a 3M^{TM} Liqui-Cel^{TM} MM-1 $\times$ 5.5 Series Membrane Contactor. The characteristics of the membrane are given in Table 1. The feed and osmotic solution were in contact with a countercurrent flow. The feed solution flowed on the lumen side and the osmotic solution was on the shell side. The weight of the feed reservoir was measured constantly using a balance (LP 4202I, VWR, Milano, Italy), and is used in Equation (1) for calculating the transmembrane flux and in Equation (2) for the mass transfer coefficient calculation. The feed solution was always kept in a closed-cap container. The feed and osmotic solutions were kept at room temperature for most experiments, except for the temperature study, in which a cooler (Corio CD-900F, Julabo, Seelbach, Germany) and a water bath (VWB2 12L, VWR, Poole, UK) were used to change the temperature in the range of 15 to 40 °C. The temperature was measured using thermocouple thermometers (2000, TME, Birmingham, UK).

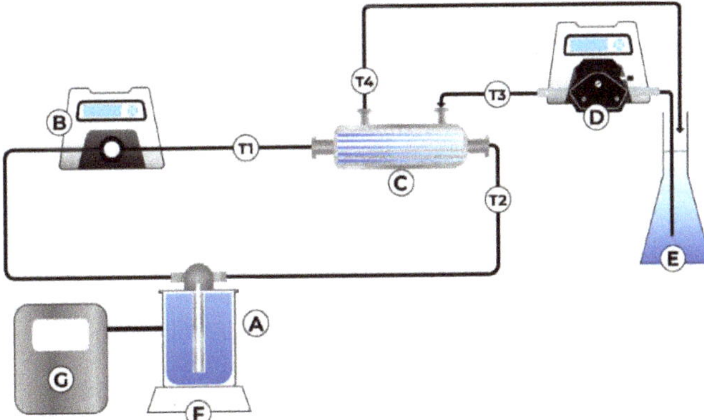

Figure 2. Schematic diagram of the membrane distillation crystallization setup: A feed solution; B gear pump; C membrane contactor; D peristaltic pump; E osmotic solution; F balance; G water bath/cooler; T1–T4 thermometers.

Table 1. Characteristics of the membrane contactor and hollow fibers.

Contactor Type	Liqui-Cel® 1 × 5.5 MiniModule™
Module configuration	Hollow fibers
Housing/potting	Polycarbonate/polyurethane
Membrane type	X50 microporous fiber
Membrane material	Pp (hydrophobic)
Porosity	40%
Effective pore size	0.04 μm
Inner diameter/outer diameter	300 μm/220 μm
Active surface area	0.18 m^2
Number of fibers	2300

Scanning electron microscopy (SEM) (GEMINI, Zeiss, Ultra 55) was used to observe the $NaHCO_3$ crystals produced at different feed temperatures. The SEM images studied were taken at 500× magnification with a signal A = E2.

X-ray diffraction (XRD) (Bruker, AXS D8 ADVANCE) was used to determine whether the feed temperature altered the crystal purity. First, a metal sputter deposition system (CEA030, Balzers, Liechtenstein) was used to coat the surface with a thin gold layer to produce a conductive surface. Subsequently, the analyses were performed with a LYNXEYE detector, with a 2Theta from 20° to 100°.

2.3. Overall Mass Transfer Coefficient and Transmembrane Flux Calculation

Two parameters allow characterization of the operating conditions of the membrane system, namely, the transmembrane flux (J, m^3/m^2·s) and the overall mass transfer coefficient (K_{ov}, m^3/m^2·Pa.s). J was calculated by measuring the weight of the feed tank over time and recorded in intervals of 20 min. The flux shown in the figures is an average of the fluxes during the experiment, calculated by Equation (1) [12,19,20].

$$J = -\frac{1}{A\rho_{wf}}\frac{dw_f}{dt} = \frac{1}{\rho_{wf}}\frac{w_{f(t_{i+1})}-w_f(t_i)}{t_{i+1}-t_i}$$

(1)

For calculating K_{ov}, the following relation is used [12,19,20]:

$$J = K_{ov}\Delta p = K_{ov}\left(p_f^* a_f - p_o^* a_o\right)$$

(2)

In this equation, p^* and a are the vapor pressure and the activity coefficient of the feed (f) and osmotic (o) sides, respectively, which were computed following the procedure described by Hamer et al. [21] and by Sandler [22] when the values of the osmotic coefficients were not found in the literature [23]. The vapor pressure (mmHg) was calculated using Antoine's equation, with the temperature T given in °C:

$$\phi = \frac{-\ln(a_w)}{vM_m M}$$

(3)

where a_w is the sum of the ions of the electrolyte ($-$), M_m is the molar mass of water (kg/mol), M is the molality (mol/kg), and a_w is the water activity.

3. Results and Discussion

3.1. Influence of the Fluid Dynamics

Improving the mass transfer is the key to having a lower required contact area and reducing capital costs. There are three resistances to mass transfer in the OMDC system: the feed boundary layer, the membrane, and the osmotic boundary layer. An increment in the velocity has a positive effect on reducing the lumen- and shell-side boundary resistances and increasing K_{ov}. Figures 3 and 4 show the flux and K_{ov} versus the change in the osmotic/feed velocities, respectively, while the velocity at the other side was set at a constant value. In addition to the boundary resistances, membrane crystallization is significantly affected by the phenomenon of concentration polarization (e.g., when the concentration of the salt is higher on the surface of the membrane), and therefore higher velocities in the membrane are more suitable, since they result in higher turbulence and thus better mixing of the solution in the membrane contactor. However, it can be observed in Figure 3a that overall, the feed velocity presents a maximum flux at 0.04 m/s when the osmotic solution operates at 0.01 m/s. At higher velocities of the feed solution, there is a decrease in flux. This decrease might be because of partial wetting of the membrane pores. As previously reported [17], wetting of the pores results in a lower flux and a higher resistance to mass transfer. In Figure 3b, with increasing feed velocity, K_{ov} decreases slightly, reinforcing the idea of potential membrane wetting. The error bars for the feed flow rate of 0.01 m/s are around 13%. Regarding the osmotic velocity, K_{ov} increases slightly when a higher velocity is used, which is an indication of more turbulence on the osmotic side and lower resistance to mass transfer. It can be observed in Figure 4a that in general, there is a rise in flux with an increase in the osmotic velocity. The maximum flux was observed at an osmotic velocity of 0.01 m/s. In Figure 4b, K_{ov} increases when the feed flow rate is higher, to overcome the resistance in the osmotic boundary layer. However, there is a drop after 0.015 m/s due to possible membrane wetting. It can be concluded that the effect of the osmotic flow rate is higher than that of the feed flow rate, and it is more favorable to have a higher osmotic flow rate than a lower feed flow rate to avoid membrane wetting.

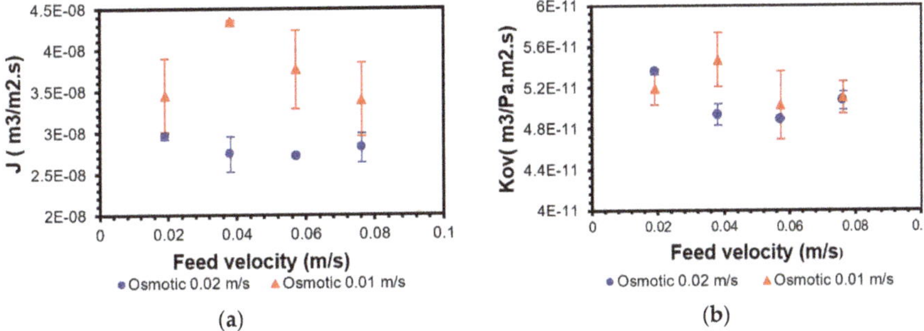

Figure 3. Effect of feed and osmotic velocities on (**a**) flux and (**b**) K_{ov}. The concentration of the $NaHCO_3$ was at a maximum, and feed and osmotic solutions were at room temperature.

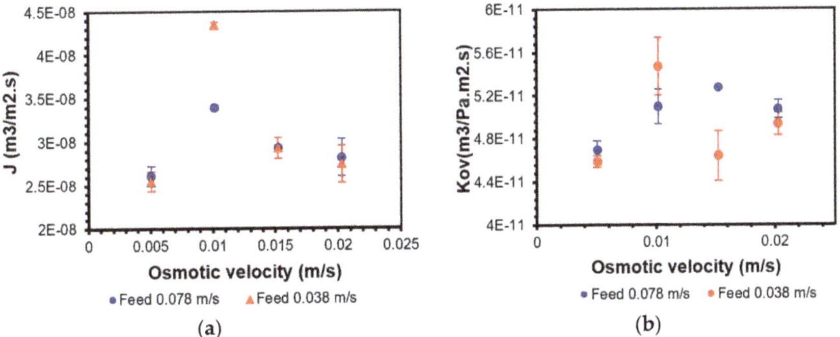

Figure 4. (**a**) Flux for changes in osmotic flow rate. (**b**) K_{ov} for changes in osmotic flow rate. The concentration of $NaHCO_3$ was at a maximum, and the feed and osmotic solutions were at room temperature.

Another set of experiments was performed to check whether there was total membrane wetting. This would be the case if NaCl was found in the feed solution. Ultrapure water was placed in the feed container, and the conductivity of the feed was measured over time with a conductivity meter. It was concluded that there was a high mass transfer of NaCl salts to the feed container at high velocity. For example, when the feed and osmotic velocities were 0.078 and 0.02 m/s, respectively, the conductivity of the ultrapure water changed from 13 μS/cm to 32 mS/cm within 2 h. This also confirms the hypothesis of partial membrane wetting at higher flow rates. Thus, velocities of 0.038 m/s (200 mL/min) for the feed side and 0.01 m/s (~200 mL/min) for the osmotic side were chosen as the optimal conditions, leading to a high K_{ov} without significant membrane wetting. These velocities were set as constant values for the rest of the experiments described in the following sections.

3.2. Influence of the Feed Concentration

Figure 5 shows the flux and K_{ov} versus the change in $NaHCO_3$ concentration. The average flux decreases with an increase in concentration. This is due to a decrease in the driving force. In osmotic membrane crystallization, the driving force for water evaporation is the vapor pressure difference between the two sides of the membrane, which is influenced by the water activity. To promote flux, the driving force must be increased. This can be achieved either by increasing the osmotic concentration or by decreasing the feed concentration. An increase in the osmotic concentration implies a lower water activity, while a decrease in the feed concentration induces a lower water activity. Globally, this results in a higher driving force [7]. Therefore, we expect to see a drop in flux with an increase in feed concentration, as can be observed in Figure 5a. By calculating K_{ov} using Equation (2), the effect of the driving force will be removed, and a constant K_{ov} is expected with a change in concentration. However, it can be observed in Figure 5b that K_{ov} still decreases with an increase in the concentration. The factor that causes K_{ov} to decrease could be concentration polarization.

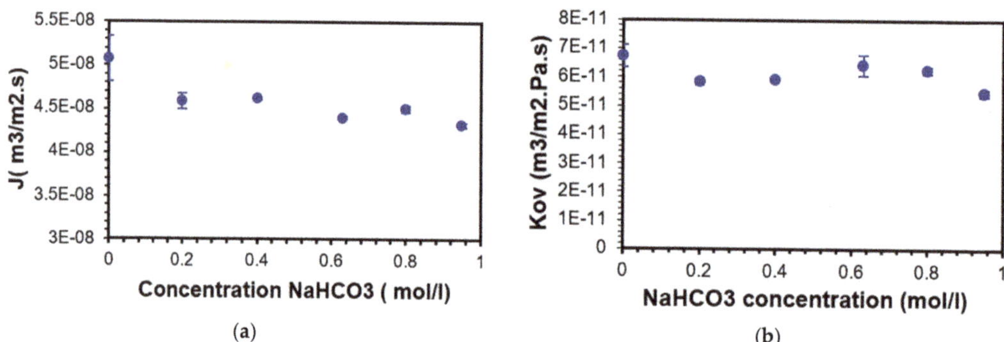

Figure 5. Effect of the concentration of $NaHCO_3$ in the feed solution on: (**a**) flux and (**b**) K_{ov}. Feed velocity was 0.038 m/s, osmotic velocity was 0.01 m/s, and osmotic and feed solutions were at room temperature.

3.3. Influence of the Feed Temperature

The temperature of the feed solution was varied in a range between 15 °C and 40 °C, while the osmotic temperature was kept at 20 °C, equivalent to room temperature (see the evolution of temperatures shown in Appendix A). This study was limited to 40° by the membrane contactor characteristics. To investigate the effect of feed temperature on the flux and the overall mass transfer coefficient, experiments were conducted with a feed concentration of 0.8 mol/L (67.2 g/L) and an osmotic concentration of 6.16 mol/L (360 g/L). The feed flow rate and the osmotic flow rate were at their optimal values for this system. The results are presented in Figure 6. When the feed temperature increases, the flux increases due to the vapor pressure difference created by the temperature difference and the concentration difference across the membrane at the same time. After some time of operation, the feed solution and the osmotic solution reach the same temperature, since the two streams are recirculated in the experimental setup. The membrane contactor acts as an excellent heat exchanger between the feed and osmotic solutions, which unfortunately is not the desired effect, since most of the energy is lost in heating the osmotic solution rather than evaporating the water in the feed. Thus, it is more efficient if a membrane with a lower heat conductivity is used. Unlike the flux, the mass transfer coefficient decreases when the temperature increases, which could be explained by the presence of the temperature polarization effect. These phenomena were also observed by Salmon et al. [18] and by Boubakry et al. [24].

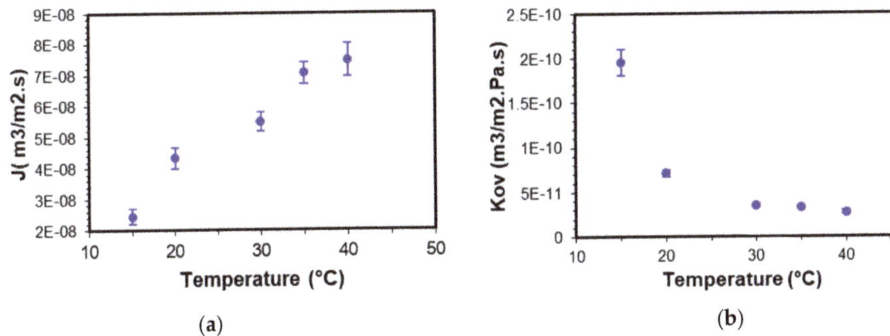

Figure 6. (a) Flux for changes in temperature, for 0,8 mol/L of $NaHCO_3$. (b) K_{ov} for changes in temperature, for 0,8 mol/L of $NaHCO_3$.

The results of previous studies on membrane crystallization are presented in Table 2, for comparison with the results obtained in this study. It can be observed that the flux obtained in this work was in good agreement with the values obtained with the same type of hollow fibers but was inferior to the flux obtained in bigger membrane contactors. The fact that most of the studies did not report the mass transfer coefficient is a critical limitation, in terms of making a fair comparison. Sparenberg et al. [20] used the same type of membrane contactor for direct contact and vacuum membrane crystallization. The vacuum membrane crystallization had a higher flux in comparison to OMCD and DCMD, as the heat losses during the process were lower. The same applied to K_{ov}: the values obtained in previous studies using the same type of hollow fibers were near the values obtained in this study, ranging from $4.8 \times 10 - 11$ m^3/m^2·Pa·s to $6.53 \times 10 - 11$ m^3/m^2·Pa·s.

Table 2. Comparison of performance of previous membrane crystallization processes with the process in this study.

Reference	Membrane Process	Membrane Type	Material	Crystal Product	J_{max} (kg/m^2·h)	K_{max} (m^3/m^2 Pa s)
[18]	OMCD	Hollow fiber	PP	$NaCO_3$	0.138	6.53×10^{-11}
[20]	VMD/DCMD	Hollow fiber	PP	$NaCO_3$	0.8 and 0.11	4.8×10^{-11} and 3.7×10^{-11}
[25]	MDC	Flat sheet	PTFE	CaCO3, NaCl, KC	6	-
[26]	MDC	Hollow fiber	PVDF/PTFE particles	NaCl	8	-
[27]	VMDC	Hollow fiber	PVDF	$Al(NO_3)$	9.6	-
[28]	MDC	Flat sheet	Elongated PTFE	NH_4NO_3	2–5	-
This study	OMCD	Hollow fiber	PP	$NaCO_3$	0.269	6.41×10^{-11}

3.4. Crystalline Phases

Commercial crystals were observed via SEM for comparison with the crystals produced at 15, 20, 30, 35, and 40 °C. The images produced via SEM are shown in Figure 7a–f. Commercial sodium bicarbonate is a powder consisting of flat sheet crystals with no preferential shape, while the crystals produced by membrane distillation crystallization were in the form of squares, sticks, and other shapes. The shape of the crystals obtained in this study was similar to that of crystals obtained in the literature using other novel crystallization processes. Therefore, the effect of OMDC on the shape of the crystals was not significant [17,29,30]. The temperature influenced the morphology and the size of the crystals, giving bigger crystals at 35 °C. This is due to the fact that temperature has a great effect on the nucleation and growth rate of crystals [16]. The experiment at 35 °C gave larger crystals. This experiment was repeated four times, and it was observed that on two occasions the crystal size was similar to the size in the experiment at 40 °C, while on the other two occasions bigger crystals were obtained. For this reason, the average size should be taken carefully. This is due to the effect that temperature, or the residence time of the crystals in the tank, has on the nucleation and growth rate of crystals. The crystal size obtained in this study agreed with the crystal size obtained by Adnan Abdel-Rahaman et al. [31]. To find an optimal temperature, higher temperatures may have to be tested, but this was not possible in this study due to the thermal limitations of the material of the module.

Figure 7. SEM images of NaHCO$_3$, comparing commercial crystals (**a**) with crystals obtained using osmotic membrane distillation crystallization at different temperatures: (**b**) 15 °C; (**c**) 20 °C; (**d**) 30 °C; (**e**) 35 °C; (**f**) 40 °C.

XRD analysis was performed on various bicarbonate crystals. The first was the original sodium bicarbonate powder from the industrial supplier. The second to the sixth samples analyzed were $NaHCO_3$ crystals obtained after membrane crystallization distillation at 15, 20, 30, 35, and 40 °C for the feed solution. A comparison of the different XRD spectra is shown in Figure 8. As previously reported [32,33], all the spectra compared showed peaks at 29.7, 35.4, and 40.8 (2Theta), attributed to the $NaHCO_3$ crystal phase.

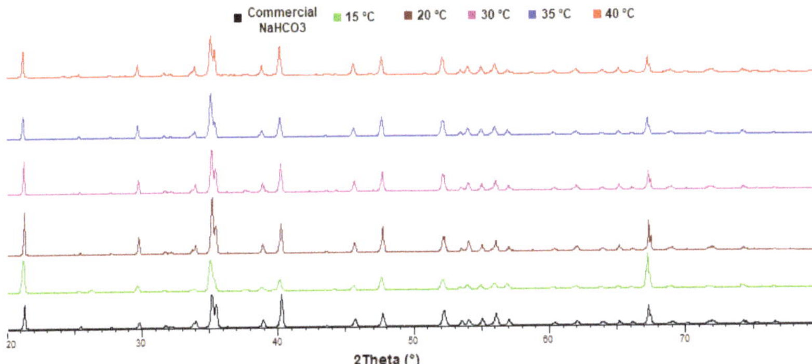

Figure 8. $NaHCO_3$ crystals—X-ray diffraction analysis.

The purity of the crystals obtained was in a range between 96.4 and 100%. While the crystals obtained at 30° had the lowest purity, showing a composition of 96.4% pure $NaHCO_3$ and 3.6% hydrated Na_2CO_3, the highest purity was observed at 20, 35, and 40 °C, with crystals of 100% $NaHCO_3$. The quantitative analysis of purity is included in Appendix B. As observed by Wang et al. [34], two factors influence the decomposition process of $NaHCO_3$ to $Na_2CO_3 + CO_2 + H_2O$: temperature and water activity. The change in water activity is sensitive to both temperature and the composition of the liquid. The variation in the purity of the crystal can be explained by the membrane system's energy losses while heating the osmotic solution. This produces a slight variation along the membrane that can induce decomposition of $NaHCO_3$ to Na_2CO_3.

4. Conclusions

Osmotic membrane distillation crystallization (OMDC) is a novel technology considered as an alternative to conventional crystallizers. OMDC has been studied for crystallization of sodium bicarbonate, due to its advantages such as lower energy and material consumption, control over the operational parameters, and larger evaporation surface area, among others. Several parameters such as the feed and osmotic velocities, feed concentration, and feed temperature were optimized. Regarding the feed and osmotic velocities, as the velocity increases the possibility of membrane wetting increases significantly. Therefore, a feed velocity of 0.078 m/s and an osmotic velocity of 0.01 m/s were chosen as the optimal conditions, which resulted in obtaining a K_{ov} of $5.4 \times 10-11$ $m^3/Pa \cdot m^2 \cdot s$. In addition, since the driving force in OMDC is the difference in concentration, an increase in feed concentration reduces the driving force and results in a reduction in the flux. However, when studying K_{ov} and removing the driving force effect, the process was found to be affected by concentration polarization, and K_{ov} still decreased by 23.6%. Finally, the effect of the temperature on water evaporation showed that the driving force of the system increased with temperature, as the flux increased from 2.45×10^{-8} to 7.49×10^{-8} $m^3/m^2 \cdot s$, but a great deal of energy was lost via the heat exchange between the feed and osmotic solutions. It was also observed that the size and the purity of the crystals were affected by the temperature, with larger sizes and higher purities obtained at higher temperatures.

5. Patents

The process presented here is registered under the patent application EP 2021163.

Author Contributions: Conceptualization, M.G.A. and V.S.S.; methodology, M.G.A. and V.S.S.; formal analysis, M.G.A. and V.S.S.; investigation, M.G.A., V.S.S., M.B. and A.C.; writing—original draft preparation, M.G.A. and V.S.S.; writing—review and editing M.G.A., V.S.S., R.B.S. and P.L.; visualization, P.L.; supervision, P.L.; funding acquisition P.L. All authors have read and agreed to the published version of the manuscript.

Funding: This research was funded by the European Research Council (ERC) as part of the European Union's Horizon 2020 research and innovation program (grant agreement ERC Starting Grant UE H2020 CO2LIFE 759630).

Institutional Review Board Statement: Not applicable.

Informed Consent Statement: Not applicable.

Data Availability Statement: Data available upon request to the authors.

Conflicts of Interest: The authors declare no conflict of interest.

Appendix A

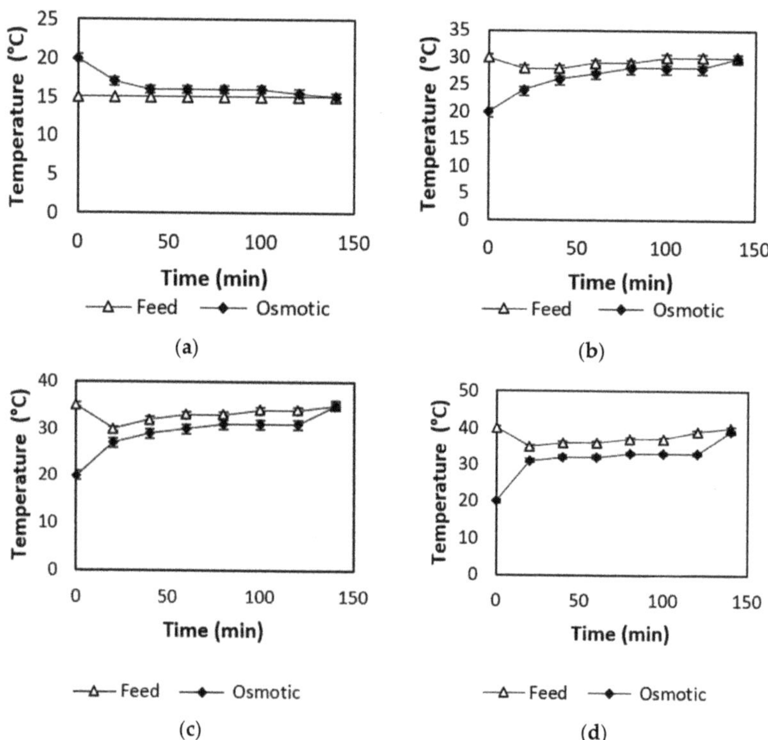

Figure A1. Feed and osmotic temperature evolution with time during the thermal evaluation of membrane distillation crystallization: (**a**) 15 °C; (**b**) 20 °C; (**c**) 35 °C; (**d**) 40 °C.

Appendix B

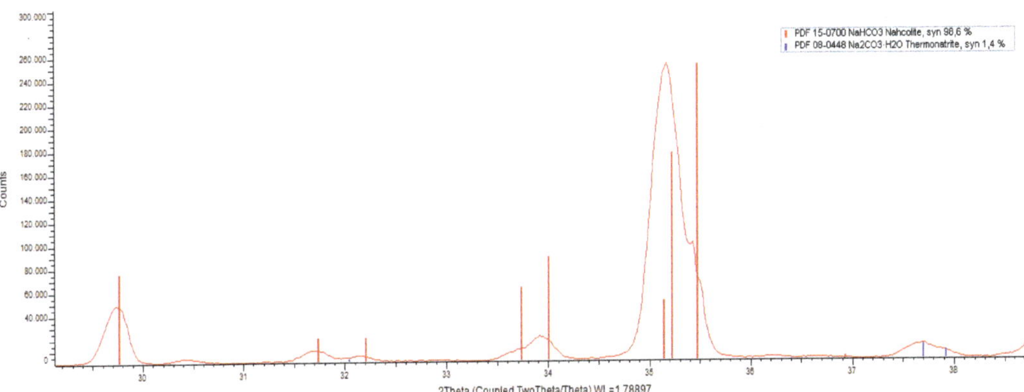

Figure A2. XRD quantitative analysis for crystals produced at 15 °C.

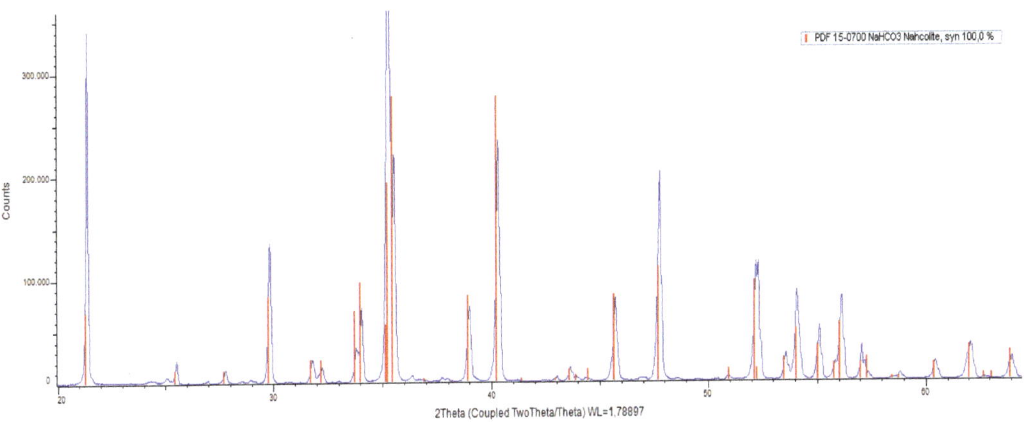

Figure A3. XRD quantitative analysis for crystals produced at 20 °C.

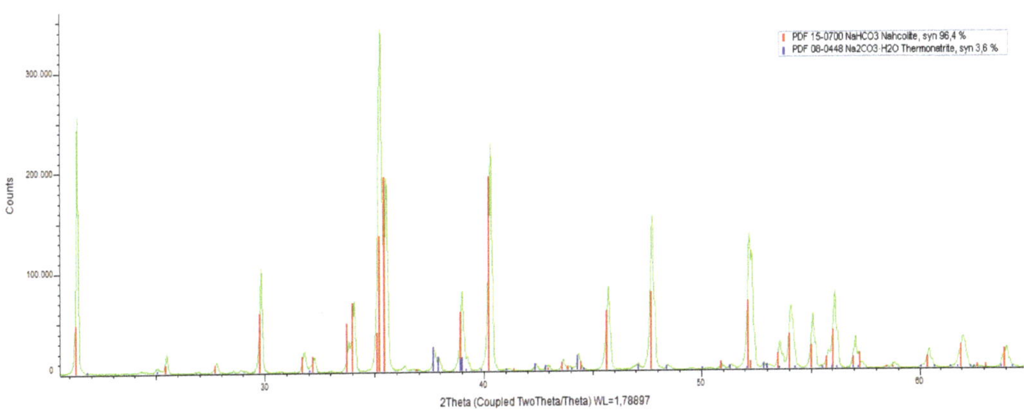

Figure A4. XRD quantitative analysis for crystals produced at 30 °C.

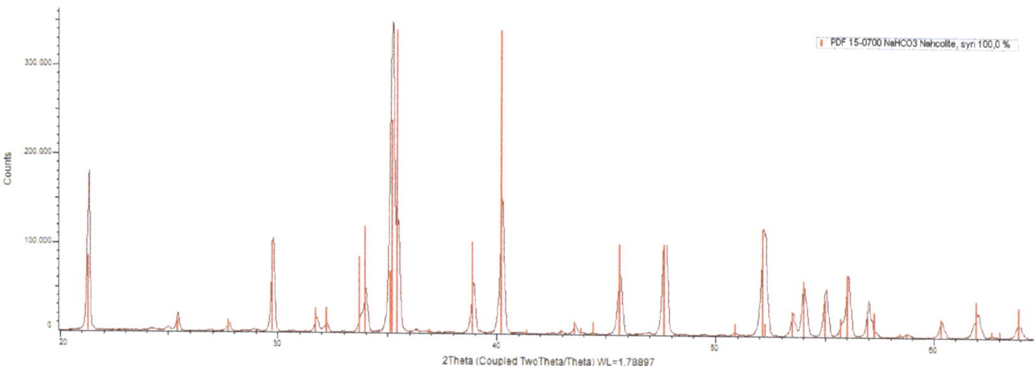

Figure A5. XRD quantitative analysis for crystals produced at 35 °C.

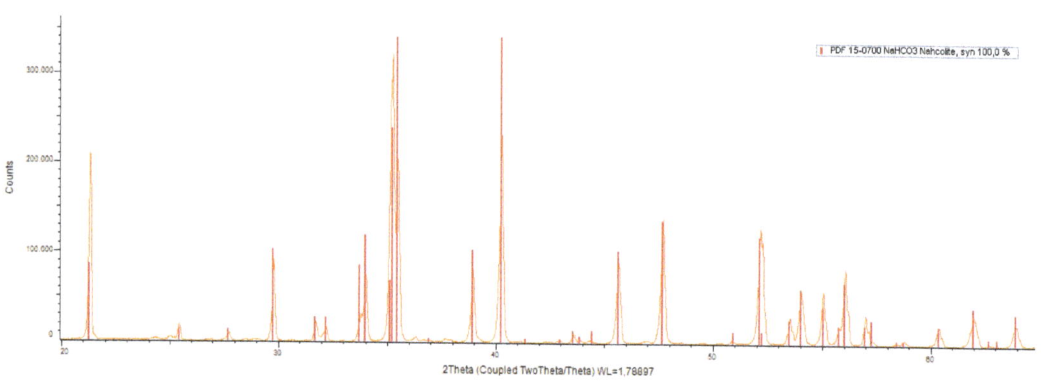

Figure A6. XRD quantitative analysis for crystals produced at 40 °C.

References

1. Edenhofer, O.; Madruga-Pichs, R.; Sokona, Y.; Minx, J.C.; Farahani, E.; Kadner, S.; Seyboth, K.; Adler, A.; Baum, I.; Brunner, S.; et al. (Eds.) *Climate Change 2014 Mitigation of Climate Change Working Group III Contribution to the Fifth Assessment Report of the Intergovernmental Panel on Climate Change*; IPCC: Cambridge, UK; New York, NY, USA, 2014.
2. Cassano, A.; Conidi, C.; Drioli, E. A Comprehensive Review of Membrane Distillation and Osmotic Distillation in Agro-Food Applications. *J. Membr. Sci. Res.* **2020**, *6*, 304–318. [CrossRef]
3. Wang, J.; Li, F.; Lakerveld, R. Process Intensification for Pharmaceutical Crystallization. *Chem. Eng. Process. Process Intensif.* **2018**, *127*, 111–126. [CrossRef]
4. Salmón, I.R.; Luis, P. Membrane Crystallization via Membrane Distillation. *Chem. Eng. Process. Process Intensif.* **2018**, *123*, 258–271. [CrossRef]
5. di Profio, G.; Salehi, S.M.; Curcio, E.; Drioli, E. Membrane Crystallization Technology. In *Comprehensive Membrane Science and Engineering*, 2nd ed.; Elsevier: Amsterdam, The Netherlands, 2017; Volume 3, pp. 297–317.
6. Drioli, E.; di Profio, G.; Curcio, E. Progress in Membrane Crystallization. *Curr. Opin. Chem. Eng.* **2012**, *1*, 178–182. [CrossRef]
7. Kramer, H.; Anisi, F.; Burak, E.; Stankiewicz, A.I. Membrane Crystallization Technology and Process Intensification. In *Comprehensive Membrane Science and Engineering*; Drioli, E., Lidietta, G., Fontananova, E., Eds.; Elsevier: Amsterdam, The Netherlands, 2017; Volume 4, pp. 1–7.
8. Chabanon, E.; Mangin, D.; Charcosset, C. Membranes and Crystallization Processes: State of the Art and Prospects. *J. Membr. Sci.* **2016**, *509*, 57–67. [CrossRef]
9. Drioli, E.A.C.; Curcio, E. *Membrane Contactors: Fundamentals, Applications and Potentialities*; Elsevier: Amsterdam, The Netherlands, 2011.
10. Drioli, E.; Curcio, E.; di Profio, G. State of the Art and Recent Progresses in Membrane Contactors. *Chem. Eng. Res. Des.* **2005**, *83*, 223–233. [CrossRef]
11. Alkhudhiri, A.; Darwish, N.; Hilal, N. Membrane Distillation: A Comprehensive Review. *Desalination* **2012**, *287*, 2–18. [CrossRef]

12. Luis, P. *Fundamental Modeling of Membrane Systems: Membrane and Process Performance*; Elsevier: Amsterdam, The Netherlands, 2018.
13. Nagy, E. *Basic Equations of Mass Transport through a Membrane Layer*, 2nd ed.; Elsevier: Amsterdam, The Netherlands, 2019.
14. Macedonio, F.; Ali, A.; Drioli, E. Membrane Distillation and Osmotic Distillation. In *Comprehensive Membrane Science and Engineering*; Elsevier: Amsterdam, The Netherlands, 2017; Volume 3, pp. 282–296.
15. Zhu, Y.; Haut, B.; Halloin, V.; Delplancke-Ogletree, M.P. Investigation of Crystallization Kinetics of Sodium Bicarbonate in a Continuous Stirred Tank Crystallizer. *J. Cryst. Growth* **2005**, *282*, 220–227. [CrossRef]
16. Saberi, A.; Goharrizi, A.S.; Ghader, S. Precipitation Kinetics of Sodium Bicarbonate in an Industrial Bubble Column Crystallizer. *Cryst. Res. Technol.* **2009**, *44*, 159–166. [CrossRef]
17. Jiang, S.; Zhang, Y.; Li, Z. A New Industrial Process of $NaHCO_3$ and Its Crystallization Kinetics by Using the Common Ion Effect of Na_2CO_3. *Chem. Eng. J.* **2019**, *360*, 740–749. [CrossRef]
18. Salmón, I.R.; Janssens, R.; Luis, P. Mass and Heat Transfer Study in Osmotic Membrane Distillation-Crystallization for CO_2 Valorization as Sodium Carbonate. *Sep. Purif. Technol.* **2017**, *176*, 173–183. [CrossRef]
19. Sparenberg, M.C.; Ruiz Salmón, I.; Luis, P. Economic Evaluation of Salt Recovery from Wastewater via Membrane Distillation-Crystallization. *Sep. Purif. Technol.* **2020**, *235*, 116075. [CrossRef]
20. Sparenberg, M.C.; Hanot, B.; Molina-Fernández, C.; Luis, P. Experimental Mass Transfer Comparison between Vacuum and Direct Contact Membrane Distillation for the Concentration of Carbonate Solutions. *Sep. Purif. Technol.* **2021**, *275*, 119193. [CrossRef]
21. Brito, M.M.; Jullok, N.; Rodríguez, Z.N.; van der Bruggen, B.; Luis, P. Membrane Crystallization for the Recovery of a Pharmaceutical Compound from Waste Streams. *Chem. Eng. Res. Des.* **2014**, *92*, 264–272. [CrossRef]
22. Sandler, S.I. *Chemical, Biochemical, and Engineering Thermodynamics*, 5th ed.; Wiley: Hoboken, NJ, USA, 2014; ISBN 1118915178/9781118915172.
23. Pitzer, K.S.; Peiper, J.C. Activity Coefficient of Aqueous Sodium Bicarbonate. *J. Phys. Chem.* **1980**, *84*, 2396–2398. [CrossRef]
24. Boubakri, A.; Hafiane, A.; al Tahar, S.B. Nitrate Removal from Aqueous Solution by Direct Contact Membrane Distillation Using Two Different Commercial Membranes. *Desalination Water Treat.* **2015**, *56*, 2723–2730. [CrossRef]
25. Creusen, R.; van Medevoort, J.; Roelands, M.; van Renesse van Duivenbode, A.; Hanemaaijer, J.H.; van Leerdam, R. Integrated Membrane Distillation–Crystallization: Process Design and Cost Estimations for Seawater Treatment and Fluxes of Single Salt Solutions. *Desalination* **2013**, *323*, 8–16. [CrossRef]
26. Luo, L.; Zhao, J.; Chung, T.S. Integration of Membrane Distillation (MD) and Solid Hollow Fiber Cooling Crystallization (SHFCC) Systems for Simultaneous Production of Water and Salt Crystals. *J. Membr. Sci.* **2018**, *564*, 905–915. [CrossRef]
27. Shi, W.; Li, T.; Tian, Y.; Li, H.; Fan, M.; Zhang, H.; Qin, X. An Innovative Hollow Fiber Vacuum Membrane Distillation-Crystallization (VMDC) Coupling Process for Dye House Effluent Separation to Reclaim Fresh Water and Salts. *J. Clean. Prod.* **2022**, *337*, 130586. [CrossRef]
28. Bush, J.A.; Vanneste, J.; Leavitt, D.; Bergida, J.; Krzmarzick, M.; Kim, S.J.; Ny, C.; Cath, T.Y. Membrane Distillation Crystallization of Ammonium Nitrate Solutions to Enable Sustainable Cold Storage: Electrical Conductivity as an in-Situ Saturation Indicator. *J. Membr. Sci.* **2021**, *631*, 119321. [CrossRef]
29. Zhu, Y.; Demilie, P.; Davoine, P.; Cartage, T.; Delplancke-Ogletree, M.P. Influence of Calcium Ions on the Crystallization of Sodium Bicarbonate. *J. Cryst. Growth* **2005**, *275*, e1333–e1339. [CrossRef]
30. Ga, R.S.; Seckler, M.M.; Witkamp, G.-J. Reactive Recrystallization of Sodium Bicarbonate. *Ind. Eng. Chem. Res.* **2005**, *44*, 4272–4283. [CrossRef]
31. Abdel-Rahman, Z.A.; Hamed, H.H.; Khalaf, F.K. Optimization of Sodium Bicarbonate Production Using Response Surface Methodology (RSM). *Diyala J. Eng. Sci.* **2018**, *11*, 22–28. [CrossRef]
32. Yoo, M.; Han, S.J.; Wee, J.H. Carbon Dioxide Capture Capacity of Sodium Hydroxide Aqueous Solution. *J. Environ. Manag.* **2013**, *114*, 512–519. [CrossRef] [PubMed]
33. Wang, Y.; Cheng, Y.-S.; Yu, M.-G.; Li, Y.; Cao, J.-L.; Zheng, L.-G.; Yi, H.-W. Methane Explosion Suppression Characteristics Based on the $NaHCO_3$/Red-Mud Composite Powders with Core-Shell Structure. *J. Hazard. Mater.* **2017**, *335*, 84–91. [CrossRef] [PubMed]
34. Wang, Q.; Li, Z. A Modified Solvay Process with Low-Temperature Calcination of $NaHCO_3$ Using Monoethanolamine: Solubility Determination and Thermodynamic Modeling. *AIChE J.* **2019**, *65*, e16701. [CrossRef]

MDPI

St. Alban-Anlage 66

4052 Basel

Switzerland

Tel. +41 61 683 77 34

Fax +41 61 302 89 18

www.mdpi.com

Energies Editorial Office

E-mail: energies@mdpi.com

www.mdpi.com/journal/energies